ON THE SCENT

ON THE SCENT

A JOURNEY THROUGH THE SCIENCE OF SMELL

PAOLO PELOSI

OXFORD
UNIVERSITY PRESS

OXFORD
UNIVERSITY PRESS

Great Clarendon Street, Oxford, OX2 6DP,
United Kingdom

Oxford University Press is a department of the University of Oxford.
It furthers the University's objective of excellence in research, scholarship,
and education by publishing worldwide. Oxford is a registered trade mark of
Oxford University Press in the UK and in certain other countries

First Edition published in 2016
Impression: 1

Published in the United States of America by Oxford University Press
198 Madison Avenue, New York, NY 10016, United States of America

British Library Cataloguing in Publication Data
Data available

Library of Congress Control Number: 2015949829

ISBN 978-0-19-871905-2

Printed in Great Britain by
Clays Ltd, St Ives plc

PREFACE

Odorants constitute a very large and surprisingly diverse family of chemicals, each with their unique identity, but all interesting and captivating, as is the assortment of smells around us which add flavour to our lives.

In this book we focus on a very large number of molecules, all diverse and unrelated by structure or chemical properties, but sharing the characteristic of being volatile and therefore capable of being carried by a breath of air to our nose, where they trigger our perception of a variety of smells.

For those to whom chemistry is not the most appealing subject I suggest approaching its study through olfaction. We have an excellent analytical laboratory in our nose. What it returns when challenged by molecules is not irksome formulas and charts, but sensations and emotions, sweet as the scents of flowers, attractive as the wafting smell of baking bread, stimulating as the pungent odour of spices. Images and memories are linked to such analytical reports, making them vivid. Just try to visualize the molecules responsible for such exciting experiences and you can immediately perceive the beauty of chemistry.

As a chemist I have always been interested in smell and therefore when as a young graduate I joined the Department of Agriculture at the University of Pisa, Italy and my mentor Carlo Galoppini suggested studying the flavour of foods, I jumped at the opportunity. Two months later I found myself at the other end of the globe in the lab of John Amoore at one of the four laboratories of the US Department of Agriculture in Berkeley, California.

Carlo Galoppini provided the initial stimulus and was able to foresee back in 1974 the value of studying olfaction, and its potential applications to food quality. During his entire career he supported and encouraged my research and after his retirement we maintained a firm friendship. Over the past few years he encouraged me in the writing of this book, which unfortunately he was not able to see published, having died in May 2014.

John Amoore introduced me to the study of the relationships between smell and molecular structure, long before biochemistry entered the field. His enthusiasm as a pioneer in the field of olfaction gave me the support and passion to venture into what was at that time a largely unexplored field. He was the first to suggest that smell is related to chemical shape and structure, a theory largely confirmed by experimental data, which were the basis for biochemical studies.[1] He passed away prematurely in 1998, but not before having witnessed tremendous changes and innovation in the field which he had greatly contributed in generating.

During my lectures to students of Food Science, and talks delivered in different contexts to general audiences, I have always experienced a lively and compelling interest in the many different aspects of smell and how we perceive it. Questions were pouring out from a wide spectrum of interests, evidencing the curiosity and yearning for know-ledge in a field where we still have difficulties in putting our everyday experiences into a scientific framework.

Smells often go unnoticed and are vastly overlooked. Yet they can secretly affect our mood, influence our choices, and make our life more enjoyable. We are surrounded by molecules continuously bom-barding our noses, even if we are not deliberately sniffing the air. Smells can be aggressive and repel us, often from dangerous situ-ations, or they can be irresistibly attractive, in stealthy ways. Smells inspire emotions and bring back with vivid immediacy long forgotten memories.

Smells are important to humans, but they are even more critical for most animals, from insects to mammals. Nevertheless the study of olfaction has been neglected until very recently. When I started my journey through smells and proteins, towards the end of the 1970s, the field was a new territory to be conquered, with very few scientists involved, and these were mainly working on psychophysical or electrophysiological aspects. Today, olfaction is one of the most active areas of neuroscience.

The study of olfaction also unexpectedly revealed another hidden treasure. The olfactory epithelium, where the endings of olfactory neurons reach the external environment—in practice an extension of the brain looking out to the world—also houses pluripotent and very primitive stem cells used in recent times to clone mice. Already by the early 1980s Pasquale Graziadei and Ariella Monti-Graziadei, an Italian scientist couple working in Florida, had amply illustrated the plasticity of the olfactory system and the unique capacity of the regeneration of olfactory neurons.[2] For these reasons this system has recently become the object of wide interest among scientists working in the field of neuronal differentiation and regeneration, as well as those studying stem cells and their uses in treating degenerative diseases.

Despite all these interesting considerations, what kept scientists away from approaching the sense of smell was its incomparable complexity. By contrast, the simple codes at the basis of colour vision and taste had been correctly guessed since ancient times from simple observations. The unique complexity of olfaction, too, is the main reason for the great difficulties, still far from being solved, in analysing and reproducing smells in ways similar to those we use for images and sounds.

Since the mid-1970s, when I first approached the field of olfaction, I have had the unique opportunity of witnessing the development of research on this sense so long ignored, from the first confusing and debated theories to the most recent achievements of molecular biology. Throughout this period, the study of olfaction has moved from psychology and psychophysics to a detailed structural analysis of the proteins and the other molecules responsible for our perception of odour, to address in recent times questions about the processing of olfactory signals in the brain, their interactions with memory, and the unique capacity of smells in stimulating us to recall past experiences and elicit vivid emotions.

At the same time, the study of olfaction has provided us with the magic power of understanding the language used by other animals to communicate. Insects, in particular, with their fascinating diversity, have developed sophisticated and accurate ways of exchanging important messages. Understanding their meanings is like putting on King Solomon's ring, as Konrad Lorenz suggested when studying the behaviour of animals.[3]

All the knowledge acquired during this time has not reduced the fascination and sense of adventure and excitement experienced by scientists working in the field. It has been both extremely interesting and rewarding for me personally and it is a passion of mine to communicate the excitement of my own experiences to anyone interested in this still mysterious and fascinating world of smells.

Finally, a note about some terms used in this book. In common speech the word *odour* carries a negative aspect, an unpleasant character, such as body odours or the odour of rotting food. However, in a scientific context, this work has a neutral meaning, referring equally to pleasant and unpleasant smells. Also, *aroma* and *flavour* are the correct scientific terms to indicate the complex sensation arising when eating food, a combination of mainly olfactory and taste notes, but including contributions related to texture, temperature, and other perceptive modalities.

ACKNOWLEDGEMENTS

This book is the product of a long and patient distillation of experiences, emotions, and relationships that have given flavour to my life as a curious scientist. I have shared this adventure with many people—students, colleagues, friends—all of whom deserve to be thanked for their contributions, even if only for being there. Above all I want to remember Carlo Galoppini, my mentor during my first years as a post-doc student, then a colleague and a friend, who prompted me to write this book and encouraged me during the early stages of writing. He supported my research through the best part of my career and it is a great sadness to me that he could not see the accomplishment of this work, his dream as much as mine. He passed away in May 2014.

John Amoore, a pioneer in olfactory research, introduced me to the world of smells back in 1975. I was fortunate to work with him, even though it was for a very short period, and benefit from his enthusiasm for science and his warm friendship. His life ended prematurely in 1998.

For the preparation of this text, I am much indebted to Krishna Persaud and Jonathan Dean. They have spent a long time reading the chapters, correcting many errors and inaccuracies, and improving the style. Krishna is a colleague and old friend, a biochemist with whom I have shared many happy moments during our research work, as well as the disappointments and defeats which are inevitable in the researcher's life. He carefully checked the text for scientific accuracy, as well as improving its presentation and contributing helpful advice and suggestions. Jonathan is a theologian, not a scientist and provided invaluable guidance on how to present unfamiliar concepts to the lay reader, interested in science, but with only a basic grounding. Being a man of letters, he greatly improved my poor English style and painstakingly corrected many errors.

Latha Menon and Jenny Nugee of Oxford University Press were my invaluable guides through the difficult and troublesome path leading from a draft to a final manuscript. This work would not have been produced without the constant support and encouragement of Latha and the careful and constant help of Jenny. Both provided me with sage advice which greatly improved both content and presentation.

I cannot name all my students who kept me on my toes during my long career with their persistent curiosity and thirst for knowledge. During the writing of the book I often reflected on the long and detailed discussions prompted by their challenging questions.

Finally, I want to thank my readers for accepting the invitation to accompany me on this journey through the world of smell. I hope that reading the book will be enjoyable as well as informative.

CONTENTS

LIST OF FIGURES

PART 1

SMELLS AND MOLECULES

MOLECULES IN THE AIR

Smells in Our Everyday Life

SURVIVAL AND PLEASURE

Are we progressively becoming unaware of the smells around us? Or perhaps just beginning, however unconsciously, to rediscover them and appreciate a new dimension in our lives? Do we really *need* our sense of smell?

Certainly we can do without it and still lead a perfectly normal life. Some of us are unable to perceive any odour, often as a result of an accident or an acute inflammation of the nasal cavity. The lack of a sense of smell does not seem to have dramatic consequences. There is also good evidence that we humans are beginning to lose our sense of smell. The reason is that for our species the ability to perceive and correctly recognize different odours does not provide an evolutionary advantage over those who have lost such capacity. In other words, individuals with a faulty olfactory system do not experience problems of survival and reproduction, so they are able to transmit their faulty genes to their offspring.

In other animals, however, olfaction is necessary for survival and for reproduction. From the most primitive worms to mammals, from insects to fish, life for the majority of animal species depends greatly on the efficiency of their olfactory system. We can also include in this view simpler organisms such as bacteria and protozoa, if we extend the concept of olfaction to the broader one of chemoreception, that is, the ability to monitor the chemical composition of the external environment.

Using our nose to explore the environment

It's hard to conjure up the image of the world from the perspective of a dog. For us humans, vision dominates among the five senses and provides the most accurate information on the external environment. We build a visual map which helps us to navigate, to recognize people and places, and to make choices. Our emotions are stimulated to a large extent by vision and sounds; our memories are visual, as are our dreams. When we describe a place, a house, a street, a scene, we use images, as these contain a lot of information.

Does the same apply to a dog? When, for example, we take a dog to a new place, instead of looking around, the first thing he does is smell. His attention is aroused by olfactory messages issuing from hidden corners, traces left by animals of the same or other species to mark their territory, warn of dangers, or as mating signals.

Would you be able to build an olfactory map and use it like a dog to make your way around the town where you live?

Shut your eyes now, plug your ears, and concentrate on smell. Alas, in our artificial and sterile world we are not likely to find much information. How different it was in Paris in the eighteenth century which Suskind described with such vitality and such immediacy on the first page of his novel *Perfume*.[1] But our modern cities lack olfactory character. This does not mean that they are cleaner; we have simply replaced the smell of excrement and rubbish, but also of baking bread, vegetables, fruits, and roast meat, with those more elusive odours of exhaust fumes from cars and lorries, less explicit warnings that our health is at risk than that of decaying foodstuffs, and therefore more easily ignored. These are the odours of our cities all over the world, differing in intensity rather than in kind.

Even today there are remote places where the smells produced by common everyday activities are not concealed and still contribute to enrich the local culture. Let us picture ourselves strolling, with our eyes shut, in a small town in south-west China. First there is the corner where a woman sits for the entire day selling big jasmine flowers with their magical scent, so elusive that no expert perfumer has yet managed to reproduce it, and which girls like wearing round their necks. A few steps further and we are overwhelmed by the penetrating and alluring aroma of star anise gradually blending with those of other spices: we are at the entrance of the market. The sudden mouth-watering smell of fried doughnuts draws us to the stall of Mrs Wang. It is summer and the foul smell of durian fruit comes in patches, heavy and unpleasant, but powerful and captivating at the same time, lingering in the hot air, like fog. People say its taste is as delicious as its odour is repulsive. Coming out of the market, another pungent stench comes from the direction of the unmistakable sign of the public lavatory. Walking further, you know that the house of your friend Mr Li, beside the old pharmacy, must be nearby. You can *see* it with your nose, smelling the typical phenolic savour of disinfectant and medicinal herbs.

Such olfactory experiences are becoming more and more rare, but sometimes, if we are attentive enough to the messages coming from our nose, we can recognize, even in so modern a city as London, typical smells that have remained unchanged for decades: fried fish emanating from cheap restaurants, urine stagnant in some dark building entrance, or the characteristic whiff of disinfectant which used to permeate the famous red telephone boxes, a smell that became as characteristic as the kiosks themselves. Or roaming through the winding lanes of sun-drenched villages in southern Italy, among whitewashed houses against a deep blue sky, we can still catch those ancient smells of tomato and oregano simmering all morning in dark kitchens, or of local vegetables cooked according to traditional recipes.

Although emotionally powerful and effective in recalling long forgotten pleasant memories, these sensations are not essential for our everyday life. Messages coming from our nose are certainly less important and reliable than clear, detailed images and sounds. On the other hand, there are circumstances in which we strongly rely upon olfactory information. When, for example, we taste some delicious food or we want to appreciate the special bouquet of an aged wine, we focus our attention on signals coming to our nose, and often we even close our eyes in order *not* to be disturbed by visual cues, just as we shut our eyes to better appreciate a concert, focusing only on the music.

In such cases, the olfactory messages (or sounds in the case of the concert) become more important than visual stimuli. How weak is the impact produced by the images of foods and drinks, however colourful and realistic, which are continuously presented by advertisements, when compared to the elusive and appealing smell of freshly baked bread when it suddenly accosts us in the street. And how dull and depressing are those plastic reproductions, however faithful in colour and detail, of sushi and tempura exhibited in the windows of all japanese restaurants.

In any case, whatever attention we pay to smell, we will never be able to build an olfactory map of our environment. But a dog, like most animals, can *see* through its nose. Smells rather than images represent its reference points. It is impossible for us to conceive how a dog can explore the world using olfactory cues. It is not just a problem of sensitivity, even if we humans are certainly among the worst performers in terms of olfactory reaction; rather, it is a better capacity for analysing and selecting odours which dogs and other animals possess when compared to humans.

This fact is supported by the observation that in most animal species the area of the brain dedicated to processing olfactory stimuli is larger than the visual area. In humans, by contrast, the importance that the brain gives to the two senses is quite the opposite.

Signals of sex, food, danger

The perception of odours is extremely important for most animal species and many aspects of life are modulated by olfactory experiences. The survival of individuals and the preservation of species depends on a well-functioning olfactory system. It is through olfactory analysis that animals discriminate good from spoiled foods and recognize toxic substances that might be present in a potential foodstuff. Being able to detect the odour of prey is important for predators, while for potential prey it is vital to be sensitive to the odour of a predator and be aware of its presence in time to escape. It is also through odours that a potential mate advertises its presence and its availability to individuals of the same species. In many cases—this is the rule with insects—recognition of a mate of the same species relies on olfactory cues, more clear and direct than visual images, in order to avoid unsuccessful mating between individuals of different species. In some species, such as ants or honey bees, a complex olfactory language modulates hierarchy relationships within a social community and helps members recognize individuals from other communities.

While for most animal species olfaction is essential for survival and reproduction, in humans this sense contributes to making life easier and more pleasant. Unlike other animals, we do not communicate through smell between ourselves, but perceiving environmental odours is an important part of our understanding of nature. The appreciation of good food is certainly mediated by olfaction and represents for us one of the most important uses of the nose.

In fact, the complexity and richness of food flavours is mainly due to their smell components. Very often it is only tiny amounts of volatile molecules reaching our nose and stimulating our receptors that makes the difference between a dish prepared by a good cook and a fast food product, or between a bottle of aged wine and a common table wine. It is because of those few molecules that we are ready to pay high prices, testifying to the importance of such hedonistic aspects.

PERFUMES, FOOD, ENVIRONMENT

In humans, smells have always been associated with pleasure. Even in the most ancient civilizations, such as Egypt, we find traces of scented substances that were used for making oneself more pleasant and acceptable, as well as a way of masking offensive odours—in the preparation of corpses for burial for example—or as special offerings to gods. The very name *perfume* derives from the expression *per fumum*, a clear reference to the fumes of incense and other aromatic herbs that were burned during religious rites.

Besides the role of olfaction in adding *flavour* and pleasure to various aspects of our lives, the perception of smells was also important, until recent times—and to some extent still—in guiding humans (as habitually for other animals) towards correct food choices. Some very objectionable odours, such as those of amines and mercaptans, which are common degradation products of proteins, are indicators of an incipient putrefaction process, warning us that those foods are not safe to eat. We still instinctively use our noses to probe the freshness of the food we are about to eat, if for example it has been kept in the fridge for too long, before we feel confident and safe. However, the safety and the quality of foods we buy in shops or consume in restaurants is assured by a chain of controlled operations and accurate analyses. Therefore, we do not usually need to rely on our nose any longer to perform such food quality analysis.

We can also reasonably assume that the pleasant olfactory notes, such as most of those which originate when we cook our foods, long ago suggested the adoption of transformation procedures, such as cooking, which makes our food more digestible and safer to eat. In fact, a strong heat treatment destroys potentially harmful micro organisms and inactivates anti-nutritional factors, making food more easily digestible. Cooking also has the effect of denaturing proteins, which consequently become more exposed to the action of the degrading enzymes of our digestive system. Such enzymes, which are proteins themselves, break down the proteins of ingested food into their basic components, the amino acids, which are thus made available for the building of new proteins needed by our body.

We can imagine a protein as a long string of amino acids linked to one another and folded into an apparently random coil; in fact, this folding is far from being casual, following precise requirements of interaction between the various groups which, in the end, determine a stable, compact, and unique structure. Heating breaks such interactions, unravelling this thread and exposing regions which were hidden in the original compact structure, and making them available to the action of degrading enzymes.

We can guess the curiosity and pleasant surprise of the first human beings who, guided by the pleasant smell of roast meat, discovered a gazelle or a wild boar trapped in a forest fire and how such experiences may have prompted them to eat cooked meat, more tasty and easy to chew than the raw meat they had previously been consuming.

Some of these olfactory notes, potent and aggressive such as those of amines that are generated during the incorrect storage of some foods, could make our dishes too unpleasant to eat even before real danger to health is present. Several practices were developed to mask such odours and improve the organoleptic quality of foods. Amines are basic compounds and can be neutralized by the use of acidic substances. The salts thus formed are by their nature not volatile (think of common table salt) and therefore unable to reach the nose and stimulate the olfactory system. It will now be clear why we put lemon, which contains citric acid, on fish or cook meats in wine, rich in tartaric acid, or vinegar, whose main component is acetic acid. The wide use of different spices also contributed to mask unpleasant odours, which, before the use of fridges, were certainly common in stored meats and other foods.

All these procedures utilizing smells to improve the quality of life developed in the past without any scientific knowledge of the underlying chemistry. But such intuitive choices were not always correct and often properties were attributed to scents which sound naive and magic today. Sometimes perfumes were believed to be endowed with therapeutic properties and during plagues fragrant substances were usually burned, not only to mask the unpleasant stench of rotting corpses, but as a means of containing the disease.

It was only with the establishment of chemistry as a systematic discipline that olfaction began to be regarded as an object of scientific investigation. Smells are carried by molecules and before the acquisition of the tools of chemistry for studying molecules, this realm of sensation remained mysterious and elusive, as the tools for its exploration were missing or inadequate.

It was therefore only at the beginning of the twentieth century that, together with the major advances in organic chemistry, the curiosity of chemists was alerted to the attractive and elusive properties of the molecules that we perceive as smell.

Ever since chemists have been able to synthesize new molecules, their curiosity has always led them instinctively to smell what they produced, a habit still common, like a cook smelling and tasting dishes while their flavours are getting richer and more complex during the cooking. This is the first analysis we perform in the laboratory: we sniff. Even when chemists have attempted to reproduce with synthetic procedures the fragrances of natural substances, olfaction is the best instrument for checking to what extent the smell produced in the reaction flask reproduces the familiar natural character of a flower or a spice.

We can therefore easily understand why, initially, the interest of organic chemists focused on the scent of flowers, first isolating and identifying the compounds responsible for their pleasant olfactory qualities.

The ambition to create smells in the lab, to manufacture sources of appealing scents, has been the drive behind large-scale industrial enterprise. From the initial curiosity to smell the products of chemical reactions, research was aimed at specific projects for reproducing natural fragrances.

There were several advantages to such a strategy. First, an economic benefit: in most cases it is much cheaper to synthesize molecules in the lab, beginning with inexpensive petroleum, rather than extracting and purifying them from natural sources. Only in a very few cases is it still more convenient to obtain these compounds from essential plant oils. These are exceptions, when the structure of the desired compound is extremely complex and difficult to synthesize. Another advantage is ecological. It is true that the chemical industry is a source of pollution, but the impact on the environment of small industries such as those synthesizing perfumes is limited and can be further reduced by adopting correct practices of disposing of waste materials. On the other hand, the chemical preparation of perfumes has saved several species of plants and animals from extinction—the sandalwood tree is a good example.

The scent produced by this tree was already highly prized in ancient Egypt and in the East, where it grows. The scented oil is extracted from the trunk and the process involves killing the tree. The sandalwood tree only starts producing its characteristic perfume after 25–30 years of growth, so it is easy to appreciate how limited the source of this substance is, and the seriousness of the damage that is done every time a tree is felled.

Animal populations have suffered as well. Some species, such as the musk deer, the musk rat, and the civet, produce special chemicals in their glands, which they utilize as pheromones. Their olfactory properties are highly appreciated in perfumery—the so-called musk scent—and for this reason have encouraged an unrestricted hunting of these species, which have been brought to the edge of extinction.

Creative perfumery

Fortunately, the increasing interest in perfumes derived from natural sources was accompanied by more active research in synthetic chemistry and its application to the production of fragrances. Generations of chemists have engaged in synthesizing thousands of new compounds specifically to study their smell. The aim was not limited to reproducing in the lab the same structures present in nature. Several components of natural perfumes exhibit complex structures, too difficult and expensive to prepare in the lab. So chemists adopted an alternative strategy: to explore the odour of other molecules, similar in certain respects to the target natural compounds, but easier and cheaper to synthesize, in order to find substitutes for the natural substance.

This approach in turn, rather than providing simple answers, stimulated a series of basic questions about the relationships between odour and molecular structure. This was the beginning of systematic research that eventually expanded into the field of biochemistry and was responsible for finally lifting the veil of magic and mystery which had surrounded the perception of smells, moving from vague and poetic descriptions to accurate measurements of molecular parameters.

This work resulted in a strong basis of knowledge for designing and synthesizing new odorants, often endowed with novel and interesting qualities. A novel form of art, the creation of perfumes, was thus provided with new tools and ingredients which vastly enlarged its potential for expression and creativity.

Chemistry and gastronomy

There is one important field which requires our chemical senses, where science and art work together to produce results that make life more enjoyable: the field of gastronomy. In the kitchen we witness the most complex chemical reactions, which produce an enormous variety of volatile compounds that make our foods pleasant and unique. Such reactions are very sensitive to small variations in the conditions applied. It is sufficient to adjust temperature, humidity, the time of cooking, or the order in which we add the different ingredients only a little to produce a markedly different aroma for the dish being prepared. From this viewpoint, cooking can be regarded as a form of art.

Just think how we can affect the taste of meats, potatoes, onions, and other foods by cooking them at different temperatures whether we make a stew, stir fry, or roast. Or how different the crust of a freshly baked loaf tastes from the rest of the bread. It is all down to a number of volatile chemicals produced through the so-called *Maillard* reaction between components of starch and proteins.

Such a concept can also be applied to the processes of the food industry, despite the high degree of technology involved. Common examples are the fermentation of cheese, the making and ageing of wine, or the preparation of fruit juices. We are aware of the large variety of cheeses, all originating from the same curd, a rather tasteless product. All the different and sometimes strong flavours are the product of bacteria or fungi which break the molecules of fats releasing short chain acids, such as butyric, valeric, capric, endowed with *cheesy* smells. The reactions taking place during fermentation and ageing of wines produce all classes of chemicals, from the fruity notes given by esters to the phenolic compounds released by oak barrels. It is in the end the fine balance of all these compounds that affects the organoleptic quality of the final product, which now to a large extent determines its price.

In fact, once the basic requirements for a safe and nutritious product are assured, the choice among the great variety of food products on offer is mainly motivated by hedonistic considerations. We generally choose the product that we like better. To please our chemical senses we are ready to accept large differences in price even between products which have similar nutritional values. Such requests from consumers have in turn stimulated the creation of accurate techniques to evaluate the sensory properties of foods, and prompted the development of basic research aimed at understanding the molecular mechanisms which regulate our perception of flavours. Research has prompted the wine industry's use of selected yeasts in the fermentation of the must, as it has been found that a lot of chemical reactions producing aromatic compounds are governed by the different enzymes released by yeast cells. To give one example, a particular enzyme, named *glycosidase*, can release floral smelling compounds from the sugar to which they are linked in the grape, a strategy often found in plants to solubilize hydrophobic substances. As a result, the choice of a particular strain of yeast, rich in glycosidases, can produce a wine with stronger floral notes.

More recently, the concept of odour quality has also been applied to the environment. It is not enough that the air we breathe should not contain toxic components; we expect to live in a pleasant environment, free from objectionable smells. Our revulsion when confronted by bad odours is probably an evolved response resulting from the association of such smells with chemicals dangerous to health, but even when this is not the case they can affect the quality of life.

SMELL AND TASTE

When we describe the aroma of foods, we frequently use the term *taste* to indicate sensations that are usually quite complex and include, to a large extent, olfactory elements. In introducing food into our mouth, taste receptors located on the surface of the tongue are immediately stimulated and send signals to the brain which we can properly define as taste. At the same time, however, a multitude of volatile components present in our food are released, also as a result of chewing, and reach the olfactory mucosa, the area at the top of our nose, where receptors for smell molecules are located, through an opening situated on the upper wall of the palate.

It is thanks to these volatile chemicals that we appreciate the richness and variety of different food aromas. We use the terms *flavour* or *aroma* to describe the complex sensory experience generated when we taste foods, which includes olfactory and taste components, but also tactile aspects (the crispiness of chips, the crunchiness of biscuits, the melting softness of chocolate), temperature detection, and other sensory elements. Above all, smells (entering our nose through the palate and often mistaken as tastes) are responsible for the great variety and subtle differences that we appreciate in our foods.

But, of course there are proper taste sensations, which we perceive with our tongue and which provide unique contributions to the aroma of foods. At this point we should perhaps clarify the differences between smell and taste both with reference to the molecules eliciting such perceptions and to the relative sensory systems.

From a purely anatomical perspective, at least in humans and other mammals, tastes are perceived with the tongue and smells through the nose. In other words, we call *tastes* the sensations coming from our tongue and *smells* those which originate in the nose. In other species, however, this definition cannot be always correctly applied. For example, the bifurcate tongue of reptiles, which is rapidly moved back and forth, is used to sample the air and carry the odours present in the environment to the organs of chemoreception. Its rhythmic movements correspond to the rhythmic sniffing of a dog exploring the environment and navigating with its olfactory map.

If we then look at insects, we cannot find tongues or noses, but tiny sensilla, elementary sensory organs, located on the antennae, mouth parts, legs, and sometimes also on the wings and other parts of the body. However, we can regard the antennae as the main olfactory organ (the equivalent of a nose), while legs and mouths are generally dedicated to taste. But can we still talk about olfaction and taste in insects, even in the absence of noses and tongues? We certainly can, if we refer to the chemical nature of the stimuli rather than to the organs of perception.

In this case, we call odorants those chemicals carried by air to the chemosensilla of insects (and to the noses of humans and terrestrial vertebrates), while taste is elicited by non-volatile compounds, often water soluble and present in the environment.

Therefore, the antennae of insects can be regarded as olfactory organs, as they catch molecules present in the environment as gases, while the legs perceive sugars and other non-volatile compounds, which are brought into physical contact, for example when a butterfly lands on a flower or a mosquito on its host's skin. It would therefore be more appropriate to talk about contact chemoreception in insects, rather than taste.

So a distinction between olfaction and taste based on the nature of chemical stimuli rather than the anatomical structures that receive them would seem clearer and more appropriate. But it isn't so. Nature is always more complex than our schemes and definitions, and every time we feel we have arranged all the phenomena in our artificial frame we come across examples that we are not able to classify.

For fish, and for aquatic organisms generally, all chemicals are carried in water, and the above definition of odours and tastes no longer makes sense. But fish, at least, possess a nose and a tongue and we can refer to these organs to distinguish again between the two chemical senses.

ANOSMIA: BLINDNESS TO ODOURS

I mentioned earlier that we as humans are progressively losing our sense of smell. In fact we rely on vision more than olfaction to explore the environment and make our choices. Therefore an impairment in detecting smells, that are of vital importance for other animals, is no longer a threat to our life and does not prevent us from transmitting faulty genes to our offspring.

As a result, olfactory defects are rapidly accumulating in the human population. The inability to detect one or more odours is similar to the phenomenon of *daltonism*, that is, the inability to perceive one of the three basic colours. The term *anosmia* defines the conditions of individuals who cannot smell. A total anosmia, that is a complete odour blindness, is very rare in the population: in very few cases it is innate and more often could be the consequence of a physical trauma or a viral disease of the olfactory cavities. On the other hand, the inability to smell one or two specific odours is extremely common among humans. Although a systematic and complete survey of the human population has never been performed, we can certainly state that the occurrence of such defects represents the normal condition for humans; being able to perceive all types of odours constitutes the exception.

I suggested that specific anosmia is the olfactory equivalent of daltonism. This parallel assumes that there should be a certain number of basic odours—by analogy with the three basic or primary colours—which might represent the elements of a complex language. In other words, every olfactory sensation would be generated by a combination of a number of elementary stimuli. Recently, molecular biology has uncovered the nature of olfactory receptors and labelled the paths along which olfactory messages travel from the nose to the brain, showing that in fact the sense of smell works on a combinatorial code. These discoveries, that will be discussed more fully in Chapters 7, 8, and 9, have provided scientific evidence for the models proposed in the past on the basis of pure observation.

However, there is a fundamental difference between the odour code and that of colours. This difference, by no means irrelevant, concerns the number of basic elements: more than 300 types of smells in humans against only three colours. This estimate is based on the number of genes coding for *olfactory receptors*, the proteins sitting on the membrane of our olfactory neurons, each responsive to a different type of smell. We will revisit the olfactory receptors and the other proteins of the olfactory system later, to understand how they might interact with the volatile molecules of odorants and send specific signals to the brain. At this stage, however, we can better analyse the phenomenon of specific anosmia and its macroscopic effects, along with the approach followed in the 1960s and 1970s, well before the perception of odours became the object of a research at the molecular level.

We noted that the phenomenon of specific anosmia is extremely common among humans. However, we are not aware of our olfactory defects until we perform specific and detailed measurements aimed at their identification. The reason why such deficiencies are so elusive lies again in the great number of our receptors or primary odours. Unlike the condition for colour vision, where the malfuction of one of the three receptors produces major and easily detectable effects, the lack of one or two of the 300 receptors does not sensibly affect our perception of smells. There will certainly be other receptors with characteristics similar enough to those of the missing elements which can pick up the smell molecules and send some sort of signal to the brain. Whether the perceived odour of a certain compound is the same in a *normal* as in the *specifically anosmic* subject is another story.

We can confidently assume that the perceived sensations will be different, but to evaluate such differences is by no means easy. If you bring a rose to a subject who is anosmic for the specific scent of roses, he or she will certainly detect a smell, but it may be like jasmine, and they will then associate this smell with the rose and will call it *the rose scent* as learned from childhood. Only well aimed and accurate measurements will detect the inability to recognize the fragrance of roses; in this case, for example, we could ask the subject to discriminate between *rose* and *jasmine* and so identify the nature and the extent of a specific anosmia.

There is an interesting anecdote relating to the circumstances in which the first specific anosmia described in the human population was uncovered. It should come as no surprise that such a discovery happened in a chemistry lab.

As the story goes, there was a young man who was using isovaleric acid, a chemical with a very strong and repulsive smell, in his experiments. This smell is very familiar and is produced by bacteria which grows on our skin, in particular on the feet. We usually associate this smell with people who do not wash regularly, but in some contexts the same smell can become acceptable and even pleasant, if it is associated with some kinds of cheese. It is exactly the same compound, in both cases produced by micro-organisms.

It is not surprising then that the young man's colleagues started complaining about the bad smell in the lab and blamed him for not taking the necessary precautions, such as working under a fume-cupboard. The poor man looked very surprised, because he honestly could not detect any offensive odour. He assured his colleagues that the compound presented only a very faint fruity aroma. It was clear that this chemist was odour-blind to isovaleric acid, but could detect the presence of *esters*, present as small impurities in the commercial product and endowed with fruity characteristics.

Such was the beginning of a search that produced many interesting results in the following years and paved the way for scientific investigation into our sense of smell. During the years that followed, research demonstrated that about 2–3 per cent of the human population is unable to smell isovaleric acid and other organic compounds of similar structure.

The potential interest of specific anosmia was clear to Marcel Guillot in the late 1940s and afterwards to other scientists, who used a systematic study of the different types of odour blindness in attempts to break the olfactory code. This indeed was the final aim and one of the biggest objectives for those working in olfaction in that period, long before the advent of molecular biology. Even now, however, we are still far from decoding the language of odours—an extremely challenging task.

The study of specific anosmias, therefore, represented a very attractive tool for identifying one by one all the different primary odours, that is—by analogy with the three primary colours—the letters of the alphabet used in the complex language of olfactory communication.

A systematic piece of research was begun in the early 1970s by a British scientist, John Amoore, who was working in Berkeley, California at one of the four large laboratories of the USDA (US Department of Agriculture).[2] Amoore, who died prematurely in 1998 at the age of 68 while still very active in research, was a real pioneer in the field of olfaction, perhaps the first who hypothesized the existence of olfactory receptors back in the early 1950s, when biochemistry was still a young discipline and the double helix of DNA had only just been discovered.

John was the first person who introduced me to the science of olfaction and I have vivid and pleasant memories of the period I spent in his lab. It was only a few months, but that experience was fundamental for my future research work. I remember John's great enthusiastiasm for his research, his ability to communicate that excitement to others, and his willingness to listen to, and help others. He was extremely honest and strict, far from accepting compromises in his work or indeed in any aspect of his life.

John Amoore's precisely focused research led to the discovery of several other types of specific anosmias and identified the first eight primary odours.[3] Unfortunately, his ambitious project for the breaking of the olfactory code through the detailed study of anosmias was never accomplished. Today, knowing as we do that olfactory receptors, and consequently primary odours—the letters of the chemical communication alphabet—are in the order of hundreds, rather than one or two dozens as first hypothesized, it is easy to appreciate the extreme challenge of such an approach.

However, the study of specific anosmias has provided a lot of interesting and useful information. For example, now we know that the occurrence of anosmias in the human population is highly variable, depending on the type of odour. Some types are extremely rare; others are quite common and may affect nearly half of the population.

This is the case with insensitivity to androstenone, a special chemical which deserves a more detailed discussion (Figure 1). For a start, the molecule is unusually large. With 19 carbon atoms and one oxygen, it reaches the maximum size for an odorant. In fact, as the size of a molecule increases, its volatily decreases until it reaches values so low that not enough molecules can reach the olfactory epithelium.

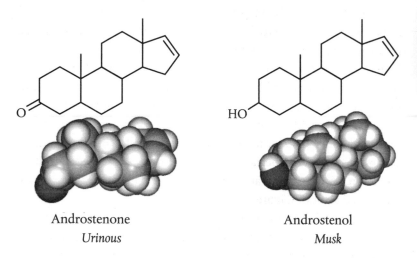

Androstenone

Urinous

Androstenol

Musk

Figure 1. Androstenone is the boar sex pheromone and strongly smells of stale urine. Its related alcohol is endowed with a pleasant musk scent. Both compounds represent the upper limit in terms of molecular size to compounds that the human nose is able to smell.

But, apart from a low volatility, there is an upper limit to the size of molecules which can interact with olfactory receptors and therefore produce an odour sensation. Our olfactory system is not able to respond to molecules that contain more than about 20 carbon atoms, because the binding sites of olfactory receptors are not large enough to accept and bind them. This second reason is probably linked to the first if we consider that larger molecules, being less volatile, would not reach the nose anyway, so, from an evolutionary perspective, why would receptors for them evolve? They would be useless and represent a waste of energy.

Androstenone belongs to the chemical class of steroids, which includes several hormones, such as testosterone, the male sex hormone. It actually comes from a molecule very similar to testosterone from which androstenone is obtained through a simple loss of a water molecule. Testosterone itself, being excreted with urine, could well have been the chemical messenger advertising the presence of a male individual, except for the fact that this molecule is not volatile and therefore is completely odourless. Androstenone, on the other hand, is volatile enough to fly to the nose and carry a very strong and specific message on the presence of testosterone, therefore of a male.

In fact, for several animal species androstenone is a love message, a sex pheromone released by the male to advertise his presence and to make the female more receptive. Often such odours are excreted with urine, together with other chemicals, which can make unique signatures. This is the case for mice and rodents, as well as other mammals. The pig represents a special case, in which the odour is conveyed by the saliva, together with another very similar chemical, androstenol, the related alcohol (Figure 1).

The blend of these two molecules constitutes the boar sex pheromone, which has a very strong effect on the sow, making her relaxed and available. The same androstenone, a strong aphrodisiac for the sow, is repulsive to us. Its odour is described as that of stale urine, at least by those of us who belong to that half of the population not fortunate enough to be anosmic to this chemical. Not being able to smell this strong, repulsive compound may have important consequences for social relationships, as the same compound can also be present in human sweat.

Androstenone can also represent a problem when detected in foods. Although it is accumulated in the salivary glands of the boar, its strong, penetrating odour can permeate the entire body of non-castrated pigs, whose meat consequently cannot be sold. For this reason, we only eat pork from young or castrated boars and of course from female pigs. In fact, the presence of androstenone in pigs is strictly related to sex and to sexual maturity. Imagine the disastrous consequences if a panel of expert tasters, required to evaluate the quality of pork, included members who were anosmic to androstenone.

Androstenol, the second component of the boar sex pheromone, is present among the volatile compounds that give truffles their typical flavour. The smell of androstenol still contains a urine characteristic, but much weaker than androstenone. In addition, its main character is that of musk.[4] This scent, which is very pleasant and appealing, is similar to that found in the glands of the musk deer and, as we saw earlier, is highly appreciated in perfumery. This molecule is probably what pigs detect when taken around for truffle hunting.

2

SMELLS AND MOLECULES

Chemical Analysis in the Nose

ODOURS ARE CARRIED BY MOLECULES

The powerful all-pervading aroma of roasting coffee, the over-whelming smell of aftershave emanating from the man sitting in front of you on the bus, the repulsive stench of an open sewer...all result from molecules floating in the air, finding their way into our nostrils and interacting with tiny anatomical structures, nerve endings poised to detect, identify, and report to the brain, in a split second.

The diversity of chemical structures translates in our minds into an impressive variety of odours. Olfaction is the easiest way to appreciate the beauty of chemistry. It is our way of 'seeing' molecules. Try to imagine molecules conveying the scent of jasmine, the flavour of a ripe fruit, sizzling bacon and fried potatoes, the bouquet of an aged wine. By contrast with flat representations of chemical structures in a textbook, the molecules suddenly come alive, brightly coloured, oxygen red, nitrogen blue, sulfur yellow, all dancing around in groups, each with their own personality—their specific smells.

But molecules are only one partner in the complex processes that generate odours. Smells are sensations, not chemical compounds.

Smells are generated when molecules interact with our nose and would not exist without an olfactory organ.

It is therefore necessary to understand how the nose *reads* the chemical messages brought by volatile molecules, and which structural aspects are decoded into the sensations we perceive. Smell is not a property of the molecule in the same way as its molecular weight or its solubility, but rather the product of the interaction between the odorant and a specific perception system.

For several decades during the second half of the twentieth century, research in olfaction was mainly focused on defining the molecular features that could best be related to specific odour characteristics. The ambitious aim of this quest was to discover the secret mechanisms used by the nose to discriminate a great number of molecules from one another, in order to break the olfactory code. A wealth of information was produced, and this gives a solid basis for extracting general rules which define relationships between odour and chemical structure.

Good smells and bad odours

Our first reaction when we come across a smell is of welcome or rejection. Before we are even consciously aware of something happening in our nose, let alone naming the smell, we have already decided whether we like it or not. The message bypasses our rational analysis and aims directly at areas of the brain that affect our instinctive behaviour.

The scents of flowers and fruits are pleasant and inebriating for everybody, those of roasted meat or freshly baked bread attractive and captivating. By contrast, the stench of urine, the putrid odour of decomposing food or that of rancid fats are instantly repulsive. In some cases, our reaction is probably innate and general and represents a sort of defence against the risk of ingesting unhealthy food or else a signal to avoid a dangerous situation.

Apart from specific cases, our reaction to smells may be strongly affected by our past experience and the links we have established between some particular types of olfactory memories and the situations in which they were generated. Thus we may like the flavour of a particular dish because it reminds us of our childhood or dislike the scent of strawberries if this was associated with a medicine we were forced to take.

The hedonic quality of a smell is also strongly affected by its strength. Olfactory images produced in our brain are extremely accurate and can elicit past memories and emotions only if they match exactly those stored in our memory and associated with past experiences. The presence of a minor component, such as a particular herb in an otherwise common dish, can immediately bring back a pleasant situation, the memory of which had been lost in some remote area of our brain.

Such an evocative effect can be so powerful that even an objectionable smell can sometimes be welcome and even perceived as pleasant when it is associated with particularly good memories. A very common example of this phenomenon is the odour of manure, which sometimes becomes acceptable as it reminds inhabitants of modern cities of a rustic and unpolluted environment.

Smells are chemical messages

We can now start following an olfactory message as it develops from the molecule carrying the chemical information to the conscious perception of the smell. First let's take a look at the structures of molecules and try to unveil their hidden olfactory properties.

Before going any further, it is important to remember that smells are elicited by molecules which have to enter our nostrils and physically interact with our olfactory receptor structures. The idea that foul-smelling compounds have already entered our nose when we perceive them may be disturbing and repulsive, but that is exactly what happens. We can relax, however, when we consider that only a small number of molecules is enough to elicit an olfactory sensation, a quantity far below the detection limit of the most advanced and sensitive analytical instruments. In fact, our nose, although a poor performer when compared with the olfactory systems of the majority of other animals, remains by far the most sensitive analytical tool available to us for detecting chemicals in the environment.

The second step is to understand how the nose recognizes the different chemical compounds. We have compared the olfactory system to a laboratory instrument, therefore the question is: what kind of chemical analysis is performed inside such an instrument?

To decode the language of olfaction, then, we need two types of complementary information: an understanding of the structures of odorant molecules, particularly those aspects related to their odours; and a knowledge of the biochemical mechanisms utilized by the olfactory system to translate the chemical information encoded in the structure of odorants into perceptions and emotions.

Information from both sides is equally important. We can draw comparisons with another sensory modality. To understand how we perceive different colours we have to possess a basic knowledge of the nature and characteristics of light, but we also need to know the range of wavelengths to which our eyes are sensitive, how many types of photoreceptors are present on the retina, what their spectral responses are, and finally how the perceived colours are related to the wavelengths of light.

In a similar way, we can approach the study of the olfactory system by looking at the molecules of odorants and trying to identify those structural parameters which are most important for the nose and which can be best related to the different odour qualities.

Questions of this type have been asked since ancient times. As olfactory perception is so immediate and involves every aspect of life, such sensations must have prompted questions and curiosity about how these experiences were generated.

The idea that volatile molecules physically interact with the structures of the olfactory organ was not generally accepted until very recently. There were earlier theories that tried to explain olfaction by assuming that the odorant molecules could emit radiation just like light and sound, and therefore stimulate our olfactory receptors from a distance. Such theories have naturally been abandoned since modern research in biochemistry and molecular biology has identified the proteins capable of recognizing odorant molecules by directly inter-acting with them. Contrasting with such fanciful theories, in the first century BC, Lucretius, in his treatise *De Rerum Natura*, based on the thought of Epicurus, had already suggested that olfactory sensations could be generated by the interactions of microscopic particles of odorous compounds with our nose.

Although the concept of molecules had not yet been developed, such a picture is not too far from our modern understanding. Lucre-tius went further in his analysis of olfaction and anticipated concepts that have been confirmed by the most recent biochemical research. In fact, he suggested that the different odour qualities could be related to the shape of such particles, those with a smooth surface generating pleasant scents, and those with rugged shapes and spokes on their surface being related to harsh and repulsive odours. It was only in the 1960s that the shape of molecules was recognized as a key factor determining scent.

In the following section, I will introduce a few simple molecules of odorants and try to extract the chemical and physical characteristics which can best be correlated with odour quality. In general, we will focus our attention on the role of stereochemical parameters, such as size and shape. Our search will be guided by familiar olfactory experi-ences, produced by chemical substances and smells which are part of our everyday experience.

SMELL AND MOLECULAR STRUCTURE

When chemists want to characterize a substance, they look first at those macroscopic properties which can easily be observed and meas-ured. For instance, our compound can be a solid, a liquid, or a gas; it can be more or less soluble in water or in organic solvents, such as alcohol or petrol; it can present a colour or be colourless; it can exhibit different types of chemical reactivity, different behaviours in relationship to temperature, and many other characteristics. All these properties are strictly related to the chemical structure and often we can predict, just by looking at the chemical structure of a substance, many characteristics, such as its physical state, its solubility, or its colour.

Can we predict the odour of molecules as well? In principle this should be possible, provided we know which molecular parameters are relevant for our nose when smelling chemical compounds. For example, can the smell of a substance be related to its physical appearance, its reactivity, or optical properties? Not at all. When we try to establish a definition of *smell*, we cannot avoid reference to the olfactory system. We can only define the *smell* as that particular property of molecules measured by the nose, and specifically by our own nose. This might seem like a trivial statement. But in fact we cannot define the smell independently from the instrument we use to measure it: our nose. This means that if we want to know the structural elements of a molecule which can be related to its odour we have no alternative but to ask our olfactory system.

On the other hand, it is clear that the same molecule is perceived differently by an insect compared with a human. It is well known that a great number of insect pheromones generate extremely powerful odours in the species which produce them with a few molecules sufficient to trigger a response in the insect. By contrast, the same compounds may be almost odourless for us. But as we have seen there are also some physiological differences between human individuals which, although minor, make the same chemical compound smell somewhat different to you compared to me.

The architecture of molecules

For those who are not familiar with chemical concepts and are not used to thinking at the molecular level, we can try to establish some basic principles to help visualize the shape and the properties of molecules.

We can for simplicity depict chemical structures using the ball-and-stick model, which may be familiar from school chemistry lessons. The balls represent the atoms and the sticks the chemical bonds that link them to each other to form the molecules. Very similar models have been used by chemists until recently not only for teaching chemistry, but also in research to help visualize the shapes of complex molecules. Today balls and sticks have been replaced by images that can be created on the screen of a computer and easily moved around, modified, and compared.

Organic compounds, representing the majority of molecules that make up a living organism, are constituted by a scaffold of carbon atoms connected to one another in a three-dimensional structure. Carbon, among the 92 natural elements occurring on earth, has the unique property of making stable bonds between atoms of itself, allowing the building of complex structures, such as linear or branched chains, rings, and three-dimensional figures. Such frameworks can be regarded as the skeletons of the molecules and represent the first step in building more complex molecules.

Occasionally other atoms can be included in this skeleton, generally oxygen, nitrogen and, less frequently, sulfur or other elements. To this framework, which we can still visualize with our ball-and-stick models, we need to add all the hydrogen atoms necessary to saturate the valences of the carbons. To build a model that can correspond to a possible molecule, it is necessary for all the valences of the atoms to be saturated, that is, all the available connections are established between the atoms. To do this, we only have to remember that a carbon atom always forms four bonds (we say that carbon has valence 4), nitrogen forms three bonds, oxygen two bonds, and hydrogen only one.

Another important factor is the size of a molecule, both in terms of the space occupied, and its weight. Often we take the *molecular weight* of a molecule as a measure of its size, as this parameter is easy to calculate, just adding up the *atomic weights* of the single atoms. To make things easier, we use relative measurements, taking the atomic weight of hydrogen, the smallest atom, as 1, and reporting those of the other atoms in relation to this reference. A carbon atom weighs the equivalent of 12 hydrogens; a nitrogen, 14; and an oxygen, 16. Therefore, we say that the atomic weight of carbon is 12, and so on. We can understand now that the contribution of hydrogens to the size of a molecule is relatively small and sometimes we prefer to simplify our model and ignore their presence.

The connections of chemical symbols for the different atoms and lines for the bonds between them represents the topology of the molecules and is enough to enable us to build a three-dimensional model. Often, particularly when we write the structures of complex molecules, we omit the symbols of the carbons and the hydrogens and even the bonds between carbons and hydrogens. Such a representation is more synthetic and just reproduces the skeleton of the molecule. However, it still contains all the information needed to build a complete model. Figure 2 shows the structures of three molecules, the first *open chain*, the second cyclic, the third aromatic, each indicated by a different symbolism and representation, from a synthetic structural notation to a *space-filling* model which reproduces quite accurately the actual shape and size of the molecule.

The first molecule is *1-decanol*; it is a long chain of 10 carbon atoms with an alcoholic group at one end. Its model looks a bit like a caterpillar and, like a caterpillar, it is very flexible and can assume many different shapes, straight, curled, or wave-like. The second structure is *cyclohexanol*; we still have an alcohol group, but this is attached to a cyclic structure of six carbon atoms. This molecule is still flexible to some extent and is usually present in the shape of a chair. The third example is an aromatic molecule, *phenol*. In this case the ring is flat and rigid, owing to a particular condition of some of the electrons used by the carbons to bind to one another. Each of the six carbons of the ring uses one electron to make a bond with the next carbon, while another contributes to a sort of cloud of charge uniformly distributed about the structure. Such compounds, which we call 'aromatic', all share a flat shape and some special chemical properties.

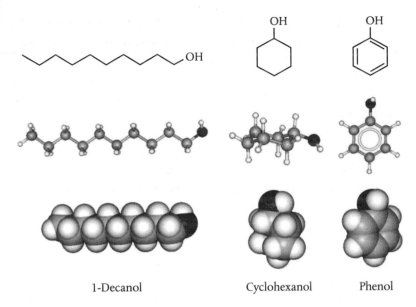

1-Decanol   Cyclohexanol   Phenol

Figure 2. Different representations of molecular structures. Top: the concise structure, where each angle represents a carbon, hydrogen atoms are omitted and only connections between major atoms are visible. Middle: the ball and stick model, with all the hydrogen atoms attached; Bottom: the space-filling model, which conveys a rather realistic idea of the size and shape of the molecule.

We can now go back to our first question: whether it is possible to predict the odour of a chemical substance merely by looking at its molecular structure. And, from the opposite perspective, can we design a molecule with a desired odour?

The short answer to both of these questions is: no. Charles Sell, a chemist making new fragrances, has been challenged many times with such a task and finally decided to summarize his conclusions in a paper whose title leaves no doubts: 'On the Unpredictability of Odours'. His arguments are based on the extreme complexity of the olfactory code, which prevents us from analysing and reproducing a smell with any accuracy.[5]

However, if we are not too demanding, and limit our task to a well-defined type of odour or group of chemical structures, we can be more optimistic and attempt to reproduce a natural smell with new molecules designed and synthesized in the lab. After all, large research groups, including Sell's, have dedicated time and resources to the synthesis of new chemicals endowed with a desired smell, but at the same time easier to make, more economical, or less toxic than that found in nature.[6]

The main problem is that the olfactory system is revealing a previously unsuspected complexity, making research in this area slow and difficult. To study such an intricate system, we need to build models that might describe situations which are oversimplified with respect to the biological system, but which allow us to obtain at least some crude and limited results.

SIMPLE ODORANTS

Already, the reader may feel we have been asking too many questions and that we risk getting trapped in a complex maze from which we shall never find our way out.

It is time to return to experience, to those common, familiar olfactory events which give, for example, zest and flavour to our morning cup of coffee, or make a pastry shop so irresistible, or let us dream, eyes closed, when sipping a glass of a particular claret. But olfactory experiences may also manifest as unpleasant signals that help us to avoid those little offerings left by a dog at the corner of the street, or warn us that our fridge badly needs a thorough cleaning.

Let us label each of those familiar experiences with chemical structures and try to *read* smells using the language of chemistry. To do this we need to oversimplify and choose some special odours and molecules. The problem is that most smells we perceive are the product of a large number of different molecules which assail our chemosensory neurons, each, to make things more difficult, eliciting complex responses. This is certainly the case for almost all the flavours we come across in our foods or in the environment, which are always the result of a great number of chemicals which together, and only in specific relative proportions, produce well-known perceptions and emotions, more or less like the sounds of different instruments in an orchestra, a sort of symphony of fragrances.

In some cases, however, and these are the most interesting and useful, the scent of a flower, or the typical flavour of a vegetable, or the environmental odour in a specific situation, are produced by single chemical compounds. We will do well to start with such examples, where the sources of the sensation (the molecules) and their relationship with the perceived smell are easier to visualize and define.

In our analysis we will bear in mind two basic concepts which will be discussed in greater detail in the following chapters, but which it will be helpful to introduce at this stage:

1. the molecules of odorants interact with specific proteins, each provided with a unique cavity—a sort of lock-and-key system;
2. to a large extent, recognition of an odorant by a protein (the fitting of the key into the relative lock) is based on stereochemical parameters: quite simply, it is the size and the shape of odorants that matters.

But before analysing the structural elements of volatile molecules and the variety of odours they can elicit in our nose, let's first consider a fact which, although obvious, is not easy to explain: even before we can consciously recognize a smell and name it, we decide whether we like it or not. It is not a rational analysis, but an immediate reaction. The same smell can be pleasant or disgusting, acceptable or repulsive, depending on the subject, on its strength or the context. Sometimes we like an odour that under different circumstances could become unwelcome; the same smell can be agreeable at low concentrations, but could be offensive when it gets too strong. Odours can also be pleasant or unpleasant depending on specific memories linked to them and evoked with unique immediacy. At this point the obvious question is: are there smells that are intrinsically regarded as good or bad by all human beings, irrespective of culture, past experience, and physiological differences?

This is not a simple question and its various aspects will be discussed in more detail later. But I will give a brief preliminary answer here. There are indeed many odours that elicit the same responses worldwide. In general, bad odours, such as spoiled food, a fire, a decaying corpse, are signals of danger; and pleasant odours—cooked food, ripe fruit, the clean air of a forest—show the way to something good for us.

It is most likely that in such cases our attitude towards certain odours has been the product of evolutionary adaptation. Those who rejected decomposing or contaminated food or escaped in time from a fire or other dangerous situations saved their lives, thus having more chances to pass their genes to their offspring. On the other hand, selecting good food and breathing clean air provided better health, longer life, and healthier progeny.

The life of other animals, from insects to mammals, depends dramatically on a correct interpretation of olfactory signals, while in humans they are mediated by culture and tradition. For us, their power has been lost in most situations, but they still produce pleasant or repulsive sensations, which act as warnings rather than compelling orders.

Offensive odours

Let's start with some nasty odours, some of which are really foul and offensive. The interesting aspect of such smells is that they are related to functional groups, such as a *carboxyl* (a carbon linked to two oxygen atoms at the end of a chain), a *thiol* (a group involving sulfur, –SH), or an *amine* (a nitrogen atom linked to one, two or three carbons), and these functional groups are the robust characters of a molecule, well-defined features, which provide direct, powerful, and unambiguous messages.

Bad odours are warnings of danger, situations to be avoided, so the information has to be very clear and convincing. These strong and immediate olfactory messages could be considered as the equivalent of sharp warnings and commands, such as STOP!, or HELP!; or like road signs which convey clear information, immediately perceived and understood without the need of additional explanation. In this respect, it is more practical and efficient if the odours carrying such messages do not change in quality across compounds of the same chemical class. This is probably the reason why, unlike the majority of odours, which are related more to molecular shape than to functional groups, these messages of danger are generally encoded in specific functional groups, which define classes of compounds typically occurring in situations to be avoided. Let's look at a few familiar examples.

Short and medium fatty acids (containing from four to 10 carbon atoms) all present the typical objectionable smell of sweat, which we have already encountered when the phenomenon of specific anosmia for isovaleric acid was described. These molecules all share a free carboxylic group at the end of the chain. Figure 3 shows the structure of isovaleric acid, as a representative member of this family, together with a description of its odour. Modifications in the hydrocarbon region of the molecule, such as the length of the chain or branching, can be responsible for additional secondary notes which may contribute to the unique odour of each fatty acid; but the main repulsive character due to the carboxyl group dominates in all these compounds.

The common names for these acids are reminiscent of their origin: butyric (butter), valeric (from the herb valerian), caproic, caprylic, and capric (from capra, the Latin name for goat). The olfactory note shared by all such compounds, that we have named *sweaty*, can also be defined as *cheesy*. In fact, all these acids are parts of the triglycerides (molecules of glycerol linked to three residues of fatty acids) that make up the fat of milk. Triglycerides are completely odourless, because of their large size and therefore reduced volatility. During the degradation of milk the free acids are generated from triglycerides by the action of enzymes which are synthesized by micro-organisms present in milk as contaminants. Therefore, the odour of these fatty acids indicates serious microbial contamination and potential danger if the milk is drunk.

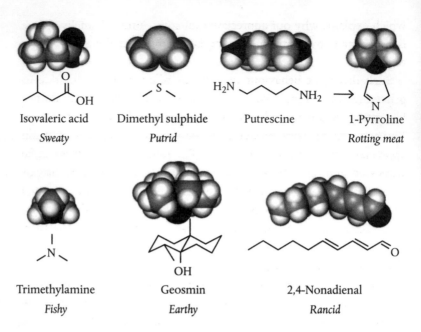

Figure 3. Examples of unpleasant odours. Generally they convey messages of danger, indicative of contamination or degradation processes occurring in foods.

But, on the other hand, we appreciate the flavour of cheese, owing to these same fatty acids. In such cases, the same flavour no longer carries a warning, but becomes a distinctive mark of quality. This is the effect of culture on the processing of olfactory messages in our brain. We are no longer alarmed by such odours if we know that the cheese we are going to eat is not the product of spontaneous degradation but was prepared following controlled techniques and the use of safe micro-organisms.

Our instinct, however, rejects such odours, which only become acceptable (and even agreeable) in the context of culture and education, through active learning. In other words, the original unpleasant sensation becomes gradually linked in our brain to positive and enjoyable experiences, becoming acceptable and reminiscent of pleasant memories. This learning process is mediated by local culture, which explains why our appreciation of some kinds of smelly cheese is not shared by many in the Far East, who find the same odour repulsive.

Such divergent behaviour in different populations is not due to genetic factors. In fact, while the majority of Chinese people do not like eating cheese, those living in the provinces of Inner Mongolia or Xinjiang, where dairy products are part of the traditional diet, do appreciate and consume such foods. The same phenomenon can be witnessed among Chinese who have spent long periods abroad or have significant contact with western cultures and who have started to include elements of foreign culture, such as cheese and wine, into their dietary habits. While many Chinese may find cheese repulsive, Europeans might feel the same about the pupae of the silk moth, which are sold as a delicacy in Seoul.

Bitter taste is another example showing how culture and education can affect our attitude and choice of aromas. Instinctively we avoid bitter foods, which we rate as unpleasant. In fact, children, without exception, do not like bitter foods or beverages. This innate attitude is also an evolutionary adaptation: bitter taste is linked to some natural substances, produced by plants, often poisonous. However, when we have enough information on a certain food to confidently think that it is safe, the curiosity and desire to explore new areas of our sensory world encourage us to try bitter products. Therefore, many of us prefer drinking coffee without sugar, many of us like bitter chocolate, or value some bitter characteristics typical of certain wines, not to mention some soft drinks and beer appreciated for their bitter character.

Another important case is the odour exhibited by small sulfur compounds, thiols, (also called mercaptans) and thioethers (which contain two hydrocarbon chains bridged by a sulfur atom). These compounds are formed during the degradation of proteins and originate from the sulfur-containing amino acids, cysteine and methionine. Being also highly volatile, they possess extremely strong and repulsive smells and, together with amines, which we will discuss shortly, are potent warnings of putrefaction processes occurring in foods. Their odour is so strong and disgusting as to produce an unconscious aversive reaction and generate an immediate sense of danger. Their efficacy in setting off an alarm signal in our brain is so great that these compounds are used as additives to cooking gas to warn us immediately of any leakage. In fact, the gases used in our stoves (methane, propane, or butane) are completely odourless, as are most hydrocarbons, and we would not be able to notice their presence in the environment without the deliberate addition of the sulfur-containing compounds.

Having stressed the repellant odour associated with the thiol group and in general with sulfur, it might come as a surprise to discover that some relatively large molecules containing sulfur are endowed with pleasant smells. It seems that when odorant molecules exceed a certain size they probably cannot fit inside the receptor protein linked to the bad smell, and acquire a different odour more dependent on the shape of the molecule. A typical example of this phenomenon is the compound responsible for the characteristic olfactory note of blackcurrants, which contains a thiol group attached to the skeleton of menthane. Another important class of pleasant-smelling compounds containing an atom of sulfur are the thiazoles, which are widely represented in the aroma of many types of foods and will be described later.

As we have already observed with isovaleric acid and other foul-smelling acids, thiols can become acceptable and even pleasant in particular contexts and at very low concentrations. In fact, dimethyl-disulphide, one of the compounds belonging to this class and very unpleasant under normal conditions, is an important component of the aroma of truffles. In the minute amounts present in this tuber it loses its aggressive character, at least for those ready to pay extremely high prices for this rare product. For many of us, however, truffles are not appetizing, and on the contrary are regarded as something to avoid because of their putrid smell.

Continuing with our description of unpleasant odours, we come to another class of repulsive compounds, the amines (Figure 3). Unlike the odours so far analysed, which can become acceptable in specific contexts, the smell of amines is always repugnant, wherever present and whatever the concentration. Amines are characterized by a nitrogen atom linked to three carbon atoms (tertiary amines), two carbons and one hydrogen (secondary amines), or a single carbon and two hydrogens (primary amines). All these molecules, provided their size does not exceed six–seven carbon atoms, have a very repulsive odour, indicating, together with the sulfur compounds mentioned above, a state of putrefaction in meat and vegetables. In fact, the degradation of proteins, in particular of two out of the 20 amino acids of which they are composed, lysine and arginine, gives rise to the diamines, appropriately named cadaverine and putrescine, respectively.

Despite their suggestive names, both these compounds are almost odourless when pure. In fact, the presence of two amine groups on the same molecule confers rather high attraction to water (hydrophilicity) and consequently low volatility. In other words, they prefer to stay in water rather than in the air. The real chemical messengers of a decomposition process assailing our noses are two cyclic compounds, 1-pyrroline (included in Figure 3) and tetrahydropyridine, produced from putrescine and cadaverine, respectively, by loss of a molecule of ammonia. Although these cyclic compounds are produced only in very small amounts, their odour is strong enough to convey an extremely powerful signal, even when the decomposition process is just beginning.

It is also an amine, in this case trimethylamine, that warns us when fish is no longer fresh. This compound is released soon after the fish dies, from the oxide, a substance of low volatility (and therefore odourless) used by salt water fish to balance the osmotic pressure.

We still make use of the offensive odour of these amines to judge the quality of meat and fish that has been stored in the fridge. It is instinctive to put these foods in front of our nose before deciding whether they are still good enough to cook. In some cases, however, although the odour is not quite right, we are confident that they are safe to eat. In fact the smell is a very early warning, and even before becoming strong and repellent can negatively affect the flavour of our dishes. We have already noted the habit of adding acid ingredients, such as lemon, vinegar, or wine, when preparing foods. This habit, besides contributing the special flavour to our dishes, also has the function of neutralizing those tiny amounts of amines, like trimethylamine or putrescine, converting them into non-volatile, and therefore odourless, salts.

One more example of a smell which normally carries a warning of danger, but which can be agreeable and welcome in other situations, is geosmin. This compound, which presents a rather complex molecular structure (Figure 3), is produced by micro-organisms called actinomycetes present in the soil. These are not active in a dry environment, but soon after a rainfall soaks the soil they wake up and start producing this substance. Geosmin is a molecule with an extremely strong smell and is responsible for the familiar and pleasant scent of wet soil which lingers in the environment after a summer storm. The distinctive note of geosmin is so highly appreciated that it is widely used in perfumery. Consequently, because it is only produced in very minute amounts by the microorganisms, much research has been dedicated to reproducing this molecule in the lab (a major task, given its complex structure), or mimicking its odour with other compounds of similar structure.

Despite its otherwise pleasant smell, geosmin can be a sign of microbial contamination, and therefore its presence becomes unwelcome when it pollutes drinking water. Being produced in the soil, it can filter into lower layers and might eventually end up in our tap water, if not adequately treated, carrying a warning signature.

The effect of concentration

To make olfaction even more complex, concentrations of odorants often have profound effects on how they are sensed. Smells which are pleasant at low levels frequently become offensive if their concentration is too high. This effect is familiar and explains why the relative concentrations of the components in a complex bouquet, like that of a wine, for example, are important for the overall perceived quality. A similar phenomenon is also encountered with visual or acoustic stimuli: the intensities of the signals and the relative strength of the components are fundamental for the message and the emotion they convey. Just think of all the balanced sounds produced by the instruments of an orchestra and how the pleasant effect can be destroyed if one instrument is played much louder than the others. Or else, consider how the same light could produce opposite effects on our mood depending on its intensity, as well as other environmental factors.

In the field of olfaction, 2,4-nonadienal (Figure 3), another molecule spelling danger, provides an interesting example of how the same compound can be present in foods at different concentrations and produce contrary emotional responses. This molecule is, together with compounds of similar structure, one of the main products in the degradation of fats. We have already seen how the majority of fats, both of plant and animal origin, are represented by triglycerides, molecules completely odourless because of their very large size. The breaking of the long fatty acid chains linked to glycerol in the molecules of triglycerides produces smaller molecules of between eight and 10 carbon atoms, including 2,4-nonadienal, which are volatile enough to reach the nose and carry the warning message. This process occurs not only in products commonly classified as fats, such as butter, oil, and lard, but also in a variety of foods that contain even small amounts of fat, such as meat, nuts, coffee, chocolate, biscuits, and many others.

Recall that in the case of milk and dairy products triglycerides can be broken down to the original glycerol and short chain fatty acids, these latter endowed with a typical sweaty-cheesy and sometimes objectionable smell. But longer fatty acids are also constituents of triglycerides: vegetable oils in particular contain only long fatty acids, generally of 16 and 18 carbon atoms. For example, olive oil is constituted by triglycerides that contain in their molecules only a few fatty acids, with oleic acid (18 carbon atoms with a double bond in the middle) accounting for about 70–85 per cent. These long-chain molecules are not very volatile and have very faint odours.

Therefore, it is the smell of breakdown compounds, such as 2,4-nonadienal and similar molecules, that alerts us to a degradation process in action. These chemicals are the final products of a chain of reactions, starting with the oxidation of a carbon atom next to the double bond by the action of free radicals and terminating with the breaking of a chemical bond around the middle point of the molecule. Therefore, if the fatty acid contains 18 carbon atoms and the double bond is around the middle, as in the case of oleic acid, the final compounds are aldehydes of eight, nine, or 10 carbon atoms. Because of their relatively small size and the nature of the aldehydic group, which is rather unwilling to interact with water (hydrophobic), these compounds are much more volatile than their parent long fatty acids. They all present similar odours, the typical familiar and unpleasant rancid note we often perceive in a bottle of oil which has been left open for several days, a piece of butter forgotten in the fridge, or a package of stale peanuts.

Amazingly, aldehydes containing nine carbon atoms, similar in structure and in odour to those rancid compounds, can be found, but in very low concentrations, in vegetables such as cucumber and, though hard to believe, also in water melon. At such low levels (we are talking about parts per billion) the pungent and disagreeable rancid smell disappears, leaving behind fresh and *green* notes.

How can the same compound produce such opposite reactions? Certainly psychology may go a long way to explaining this. Whether an odour sensation is registered as pleasant or unpleasant depends very much on how our brain processes sensory inputs, how we associate that particular smell with good or bad memories, and the meaning we attach to specific olfactory experiences. We have seen how the same smell can be repulsive or attractive, depending on the context, just as we can attach different meanings to the same word, depending on the context of the sentence. Isovaleric acid is repulsive in the sweat of an unwashed individual, but could become delicious in a mature cheese.

But there is more to explaining why odours change with concentration: something related to the 'hardware' of our olfactory system, something physiological and biochemical. Let's try to visualize the interactions of an odorant molecule with its target olfactory receptors. In general, we should assume that the same compound could be recognized by more than one receptor, generating from each of them signals of different intensities. If we recall the crude example of keys and locks, it would be as if the same key could open more than one lock, but not with the same efficiency.

Imagine you are given a key and asked to open as many locks as possible, the locks appearing in several different types. In the end, most of the locks opened would be those where the key fits nicely and without effort, but there will be other locks where the key can still work, but only after several tries and with great effort.

Returning to molecules and receptors, let us assume for simplicity that only two receptors, which we might label as *rancid* and *green*, are able to interact with nonadienal, but require different concentrations of the odorant to be activated. In particular, the *rancid* receptor could detect the compound only at relatively high levels, while the *green* receptor would be much more sensitive. A model of this type can account for the fact that the repulsive character becomes noticeable only at high levels, while lower concentrations would only activate the *green* receptor, producing a pleasant fresh sensation.

But the picture is not yet complete and there could be other phenomena involved. When we detect the nasty rancid smell, the fresh green note is not perceptible any more, even if the *green* receptor, according to our model, should be more sensitive than the *rancid* one. It is possible that bad odours, important signals of danger, are selectively amplified by the brain, which at the same time might reduce other interfering sensations. This mechanism of selective amplification has not yet been demonstrated at the physiological level, but our daily experience provides plenty of support for such an idea. We often detect unpleasant odours in our food, such as the amines in meat or fish and, although weaker in intensity with respect to the other notes, they are able to destroy the overall flavour of our dish. It is very difficult to cancel an objectionable odour by masking it with pleasant perfumes. Deodorants that are supposed to make the air fragrant by superimposing strong flowery scents on less pleasant ones usually fail in their task. It is also difficult to correct the unpleasant flavour of some food by covering it with spices.

Such differential amplification of sensory signals can also be appreciated if we consider another of our senses, hearing. Often, in the presence of a noisy background, such as that of heavy traffic, loud music in a disco, or several people speaking at the same time over the dinner table, we are able to concentrate on specific sounds which are important for us, such as particular words, or music, or a telephone ringing. We can also think of the effect produced on our perception of a single note out of tune in an orchestra. In these cases we unconsciously pay more attention and selectively *amplify* specific signals.

Returning to the odour of cucumbers, bear in mind that not everyone loves this vegetable. For some of us its odour is repulsive and the presence of a single slice of cucumber can spoil the whole salad. In these individuals the *rancid* receptor is probably more sensitive than in the average population and can be activated with the low concentrations of nonadienal and other aldehydes found in the aroma of cucumber.

Pleasant odours

So far we have looked at some special cases of unpleasant smells, produced by single chemical substances. Recall that, unlike the majority of odours, in the cases reported above, smell is mainly related to a particular chemical group, such as a carboxylic acid or an amine, rather than to the overall shape of the molecule. The presence of these groups produces types of smell which are clear and robust, not changing much with minor modification of the chemical structure, and we have related this property to the fact that such odours are usually messengers of danger.

There are also examples of pleasant smells due to a single chemical compound. We shall meet many on our trail through structures and smells, although they still remain the exceptions. Here we will look at four examples with very familiar scents—two simple alcohols, and two aromatic compounds (Figure 4). The first has only six carbon atoms, cis-3-hexenol, and a very distinctive smell of freshly mown grass. The other alcohol, slightly larger with eight carbon atoms, 1-octen-3-ol, is the typical natural odorant of mushrooms. It is astonishing how strong the evocative power of this molecule is, which alone can conjure up a vision of mushrooms under a carpet of dead leaves in a wood. If you can get hold of this chemical, you can easily improve any cheap sauce by adding just a tiny drop.

The bell pepper odorant 2-isobutyl-3-methoxypyrazine, and eugenol, which gives cloves their powerful, evocative smell, represent other examples of familiar odours generated by single pure chemicals. In all these cases, as in many others, these same compounds are responsible for the typical flavours of the natural products.

The pepper odorant 2-isobutyl-3-methoxypyrazine has several curious characteristics. This molecule is synthesized by the bell pepper plant and has often been the centre of attention throughout the short history of olfactory research for being one of the most potent known odorants. In fact, its olfactory threshold (the lowest concentration that the average human nose can detect) is as low as a few parts per thousand billion. This corresponds to a few milligrams, a tiny drop, dissolved in 1000 cubic metres of water, about the size of a competition swimming pool.

Given such a powerful smell, the amount of the pyrazine produced in the pepper is extremely small and its identification was no easy task. It was discovered towards the end of the 1960s by Ron Buttery, a scientist working in Berkeley, California, at one of the four large laboratories of the US Department of Agriculture. This same scientist later also discovered geosmin, the powerful earthy odorant that we

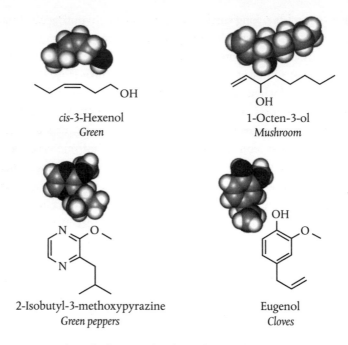

cis-3-Hexenol
Green

1-Octen-3-ol
Mushroom

2-Isobutyl-3-methoxypyrazine
Green peppers

Eugenol
Cloves

Figure 4. Examples of pleasant, familiar odours related to single chemical compounds. The two alcohols, cis-3-hexenol and 1-octen-3-ol, despite their chemical similarity, smell very different: the first is the typical 'green' odour of newly mown grass, the second is clearly 'mushroom'. The other two molecules are both aromatic compounds, but again quite different in structure and smell. 2-Isobutyl-3-methoxypyrazine is the sole volatile chemical giving bell peppers their typical flavour, while eugenol is spicy and is the odour of cloves.

came across earlier (Figure 3). The identification of these compounds, present in their biological sources in extremely small amounts, was made possible only thanks to the use of a mass spectrometer combined with a gas-chromatograph. Such a technique, now very common in all labs and essential for any scientist interested in smell research, was at that time relatively new, and Ron Buttery (who, incidentally, was working in the lab next to John Amoore's, where I was undertaking post-doc research in 1975) had built his own machine by putting together scrap parts.

Because of its strong smell, it was reasonably hypothesized that this pyrazine compound could bind in a very strong and unique way to a specific olfactory receptor protein. Therefore, it could represent a suitable *bait* to fish out the corresponding receptor from the complex soup obtained when we extract proteins from a tissue. In fact, many labs over a period of several years have used this compound for biochemical research. Such an approach was not successful in identifying olfactory receptors, for reasons that will become clear later on, but it did produce another interesting discovery: a new class of proteins able to bind and recognize different odours and pheromones. The details of this story will have to wait for a later chapter.

A story from my personal experience should convince anyone who is not familiar with this chemical how similar the scent of 2-isobutyl-3-methoxypyrazine is to that of fresh peppers. Having synthesized relatively large amounts of this compound for my biochemical experiments, my clothes had absorbed the smell, but my nose had long become used to it, so as not to detect it any longer. Although I was not aware of the odour cloud surrounding my body, I was perceived by others as a huge walking bell pepper. One evening, when I was sitting on the bus on my way home, some curious women started asking each other who had bought peppers. It was wintertime, and in those days it was not so common to find vegetables out of season. I kept silent, took a book out of my bag, and tried to hide my face behind the pages until I reached my stop, although I could not hide the smell I was carrying. At least that smell was not unpleasant, and perhaps even welcome as a whiff of summer on the wet winter evening.

It was different for the students of Leopold Ruzicka, who at the beginning of the twentieth century were busy synthesizing steroids, including androstenone, the foul-smelling compound produced in the saliva of boars which we encountered at the end of the first chapter. Later, in 1939, his research won Ruzicka the Nobel Prize for Chemistry (that he shared with Adolf Butenandt, whose name is remembered more for his discovery of the first insect pheromone than for the work on steroids for which the Nobel Prize was awarded) but at that time the population of Zurich could not see beyond a smell of stale urine, and his students were banned from all public transport.

Another example, eugenol, is the unique source of the distinctive smell of cloves. This is also a very powerful odorant, being detectable at concentrations of a few milligrams in 10,000 litres, the quantity needed to fill a large wine barrel. The wine is not mentioned by accident. Eugenol and similar compounds strongly contribute to the *woody* note of wines aged in wooden barrels, from which these compounds are released as breakdown products of lignin.

At this point we have sampled a few typical smells among the thousands filling our environment and adding character to different situations. Pleasant and repulsive, fresh and stale, food flavours, the scent of flowers, herbs and spices, arising from both nature or human activities—all are distinctive and reminiscent of specific situations, able to trigger emotional reactions and bring back vivid memories. In Chapter 3 we will attempt to organize all these smells and their related chemical structures into a sort of map which may help us to navigate our way through this intricate maze.

SNIFFING OUR WAY AROUND

A Walk Among Smells

AN OLFACTORY MAP OF CHEMICAL STRUCTURES

In Chapter 2 we visited some familiar smells and got acquainted with their chemical structures. Our journey among molecules and odours has just begun, but we already risk losing our bearings among different smells and strange molecules. To continue our exploration through this dark forest we badly need a map and a guide to show the way to the different types of smells: floral, fruity, balsamic, musk, woody, minty, and many others. With a map we can find out where we are and what comes next in the direction we are going. If we smell rancidity, how do we make the smell fresher? Should we reduce or increase the length of the carbon chain? How far is camphor from mint and turpentine? Is it really possible to build an aroma map? We can try to fix reference points and draw guidelines, but the olfactory world is too complex to be described by a map we can draw on a sheet of paper.

We first have to establish relationships between similar smells and relate such closeness to similarity in chemical structure. At the same time, we can ideally modify the structure of an odorant by adding, removing, or changing single elements and see what we get in terms of odour. Thus, we can walk through alleys and cross squares of an imaginary city, where streets and house numbers are identified by chemical families, functional groups, and structural features.

With a bit of imagination we can try to build our map of odorant molecules. It is certainly going to be an oversimplified and poorly detailed map, but it can give us some directions and a feel for our whereabouts. Crossing the streets of this chemical city, we can imagine smelling different odours and find our way around guided solely by our nose.

To be easily visualized, such a map can only be developed in two dimensions: we can go north and south, or east and west. But to describe molecules we need more dimensions. There are three elements of molecular structure that are most important with regard to odours: size, shape, and the position of the functional group. But only size can be simply defined with a number, for instance the molecular weight. When we come to shape, we have linear molecules, branched chains, flat rings as in aromatic compounds, twisted or bent rings, double or triple rings arranged in three-dimensional scaffoldings, each with several possibilities of putting side chains in different positions. How can we quantify all such characteristics?

These are only some of the difficulties we have to face if we want to put odours on a map, and they are already enough to discourage any attempt. Nevertheless, let's try by making some wild oversimplifications. Just take as reference points three types of situation: open-chain molecules (linear and branched), rings (including flat aromatic compounds and not quite flat ones), and three-dimensional structures. Strictly speaking, all molecules except aromatic compounds present a three-dimensional arrangement, but let's ignore details for now and consider our odorant molecules in basic terms.

We can set out on our journey with some simple molecules and *move* in different directions to see what type of smell we find. Our virtual promenade means changing the molecular features one by one, for example altering the type of functional group, its position on the molecular skeleton, increasing of the length of a chain, or else closing a ring or expanding the molecule in its third dimension.

We have already seen that the odour of small molecules is mainly determined by the nature of the functional group: amines have a decaying smell, sulfur compounds are putrid, carboxylic acids smell of sweat, while the small members of the alcohol family present weak, indistinct odours. We can leave these *special odours* outside our city, they are the outliers, like remains from an older settlement, which do not fit into the plan of the city.

Beginning with alcohols, we observe that only when the length of the carbon chain reaches six does a typical *green* note appear. In fact, alcohols of up to five carbon atoms only have faint, uncharacteristic odours. But suddenly we find in *cis*-3-hexenol the typical note of freshly cut grass, which we have already encountered (Figure 4), and we can imagine a garden with a green lawn. It is a very characteristic note, immediately recognizable and easily distinguishable from other *green* smells. You only start getting this scent when the grass is cut, because the chemical is synthesized on the spot, as soon as the cells are broken and all their contents mixed together. This allows specific enzymes, otherwise kept in isolated parts of the cell, to come into contact with some precursor chemicals, thus producing the volatile *cis*-3-hexenol.

What is the use of such a complicated process? Is it just to let us enjoy the keen odour of freshly mown grass? We still do not have a definite answer, but such phenomena are not uncommon. When a plant suffers an injury: mechanical damage, grazing of cattle or insect bites, it synthesizes chemicals to counteract the damage produced, such as resin to cover a wound or poison or a bitter substance to make itself unpalatable, but it also produces volatile compounds. These chemicals are now regarded as potential messengers of danger for other plants. Such an idea would have been considered as science fiction until a few years ago. Now, more and more evidence is accumulating to support the existence of chemical communication between plants mediated by volatile compounds. If you like, we might say that plants also have a sense of smell, although the word *sense* does not sound quite appropriate in organisms without a nervous system. The biochemical machinery to detect and recognize such chemical messages has not yet been studied, but it is not unlikely that plants could communicate with chemicals, just as all other organisms do, from bacteria to fungi and animals.

We then proceed along the street, increasing the number of carbon atoms, but retaining the alcohol group at the end of the chain: the odour is still green in character, but a citrus note appears, particularly in members with between nine and 10 carbon atoms. Make the chain even longer and a floral scent becomes noticeable in undecanol and dodecanol. Then, with further additions to the chain, the odour gets weaker and less characteristic. So let's go back and take a parallel street where we meet the family of aldehydes. Here, we discover odours more or less similar to those of the corresponding alcohols: green with hexanal and heptanal, a citrus character in octanal and pungent, rancid notes in the members with nine and 10 carbon atoms, chemicals which we have already found as breakdown products of fats (Figure 3). In another parallel street we could take a look at the corresponding carboxylic acids. In these compounds the characteristic unpleasant sweaty-cheesy odour of the first members (butyric and valeric acids) mingles with green and rancid notes when we reach the higher members of the family.

Once again we go back to octanol and this time we take a side street: while keeping the same framework of the molecule, we move the alcohol group to position 3 of the chain. This chemical group is important for a good interaction of the odorant with the receptor protein, as the only one able to establish a relatively strong hydrogen bond. This modification markedly affects the smell which in this case becomes mushroom-like. Add a double bond in position 1 and we are back to that special compound, 1-octen-3-ol, we remarked above as the natural one responsible for the characteristic odour of mushrooms (Figure 4). We can also try to change the functional group, from alcohol to ketone. This only requires an oxidation, which occurs spontaneously when we let mushrooms dry in the air: the odour does not change much; dried mushrooms still smell like mushrooms.

We can now compare two of the molecules we have met in this first part of our sightseeing tour: the *grass alcohol cis*-3-hexenol and the *mushroom alcohol* 1-octen-3-ol. Both belong to the same chemical class, that of alcohols, but smell completely different. The important structural difference in this case is the position of the alcohol group – OH. If we imagine how these molecules would interact with the complementary binding cavities of their specific receptors, we can reasonably assume, based on some fundamental concepts of chemistry, that it is the alcohol group that is responsible for hooking onto the receptor, because this is the only part of the molecule capable of establishing a fairly strong bond. Consequently, the two molecules would interact with their receptors in different orientations. The *oriented profile* of the grass odorant is more linear, having its functional group at one end of the molecule, and might fit into a sort of tunnel, while the mushroom alcohol needs space on both sides of the functional group in order to optimize its interactions inside the binding pocket.

We could also compare 1-octen-3-ol, the mushroom odorant, with geosmin (Figure 3), the earthy smell. The two molecules look quite different when we draw their structures. However, in terms of *oriented profiles* they share something: in both cases, the *hook*, the alcoholic group making the strongest link to the receptor, is more or less in the middle of the molecule. We also have hydrophobic regions on both sides of the –OH group in the two odorants. This similarity becomes more evident when looking at the *space-filling* models of the two molecules, which give us a better idea of their actual shapes. Would we expect their odours to be similar? In fact, they are not so different from each other and share a character that we can describe as *mouldy*.

If we increase the size of the molecule and look for odorants of 10 carbon atoms, we find ourselves in a very crowded and busy area of our town. We can imagine there is a garden or a park here, because we meet a large number of natural interesting chemicals, endowed with pleasant scents. All these molecules share a common skeleton of eight carbon atoms in a chain, with two more carbons sticking out, in positions 3 and 7, and usually a functional group on the first carbon.

These compounds are known under the common name of terpenes or terpenoids, depending on whether they are just hydrocarbons, or contain functional groups. We can look at a few examples, such as geraniol and linalool, both floral odorants or citronellol and citral: citrus-smelling. Their structures and those of the other terpenoids we are going to introduce, are illustrated together with their molecular models, in Figure 5.

These are all natural compounds, responsible for the characteristic scent of flowers, citrus leaves, and several other pleasant notes. Menthol and menthone also belong to this family, although they

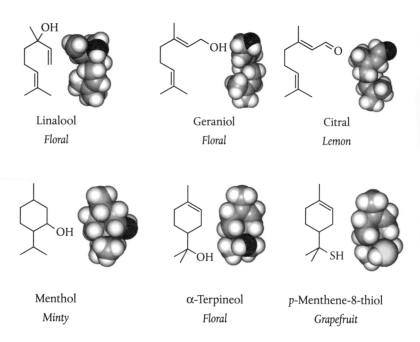

| Linalool | Geraniol | Citral |
| *Floral* | *Floral* | *Lemon* |

| Menthol | α-Terpineol | p-Menthene-8-thiol |
| *Minty* | *Floral* | *Grapefruit* |

Figure 5. Examples of terpenoids responsible for the fresh, pleasant odours of many plants. Floral, minty, and fruity notes are generally present in these volatile compounds, each of them, however, endowed with its single typical character.

appear quite different in shape. But we only need to add a single bond connecting carbons 1 and 6, to obtain their cyclic structures from those of citronellol and similar compounds. The shape of these cyclic compounds is not flat, but can be imagined as that of an elongated chair. In fact, in chemistry we speak about *chair* conformations when we draw the structures of these molecules. Menthol and menthone, needless to say, are endowed with the strong and unmistakable smell of mint and represent the main constituents of the volatile compounds of mint leaves. However, other compounds, such as pulegone, piperitone (from the botanical name of mint, *Mentha piperita*), and others, contribute to making the natural bouquet of different species of mint more complex, rich, and characteristic, just as a musical note acquires richness and character when accompanied by its higher harmonics.

The number of terpenoids occurring in nature is countless and each is endowed with its typical unique olfactory character, although all of them share some basic features. Let's now introduce a small variation on the molecule of menthol and see what we get. If we *move* the −OH group to position 8 of the molecule, we get the structure of α-terpineol, an interesting odorant with a floral smell. This is further evidence of the link between the open-chain odorants, such as linalool, and cyclic ones.

Two other derivatives bearing an −SH group on the same skeleton do not smell bad at all, as we might have wrongly deduced from the putrid stench of small mercaptans (see Figure 3). In fact, p-menthen-8-thiol smells of grapefruit, while adding a ketone group in position 3 of this molecule produces the strong and typical note of blackcurrants. Again, the shape of the molecule, rather than the type of functional group, plays the major role in determining its odour.

From mint to camphor is not a long way. These two notes are found together in the smell of several plants, and sometimes they are mistaken for one another when people are asked to recognize and name smells. So we only have to walk around the corner to stumble upon a variety of interesting chemical structures. They

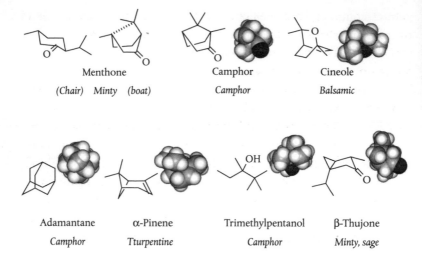

Menthone

(Chair)   Minty   (boat)

Camphor

Camphor

Cineole

Balsamic

Adamantane

Camphor

α-Pinene

Tturpentine

Trimethylpentanol

Camphor

β-Thujone

Minty, sage

Figure 6. Camphor smelling compounds are unique as they present an almost round shape. Camphor itself can be visualized like that derived from menthone by introducing an additional bond between two carbon atoms. Other compounds endowed with this odour belong to different chemical classes and present apparently different structures. However, they all share with camphor a medium size and a spherical shape.

look like strange architectural experiments, but if we take a closer look we can immediately discover how similar they are to menthol and the other terpenoids we have examined. In fact, we can easily convert menthone into an isomer of camphor (the difference is only in the position of the carbonyl group) by connecting two carbon atoms with a single bond. In Figure 6 menthone is depicted in its *boat* conformation, a form less stable than the *chair*, but still present and more suitable in our case for visualizing the similarities between the two odorants.

We have thus obtained a molecule that is almost spherical, as can be easily seen from the model. Given this highly symmetrical structure, the position of the functional group (the oxygen linked to the ring with a double bond) becomes irrelevant, since on the surface of a sphere all points are equivalent. Consequently, we find a large range of natural compounds of medium size (about 10 carbon atoms) and round shape sharing with camphor its typical scent.

Common examples are cineole (also called eucalyptol, obviously the main odorous compound of eucalyptus leaves) and fenchone, both very similar to camphor in their structures. What is more surprising—but now reasonably explained on the basis of what we have learned so far—is that other synthetic chemicals, although unrelated to camphor, like adamantane, or even open-chain compounds, such as trimethylpentanol, also exhibit similar smells. The molecule of adamantane presents a structure which, repeated indefinitely in three dimensions, produces the scaffolding by which carbon atoms are joined to one another in the structure of diamond. You can rotate the molecule of adamantane by 90 degrees in any direction, as you can with a crystal of diamond, and it looks the same.

In other natural and synthetic chemicals, the camphor character may be accompanied by other notes, such as in the two isomers of pinene, named alpha and beta. One of them is represented in Figure 6; the other only differs in the position of the double bond. The scent of these chemicals is described as 'turpentine', a note that we classify as similar to camphor, but not identical. The two pinenes are rare examples of hydrocarbons endowed with a pleasant and distinctive smell. Their names betray their origin, the pine resin of which these two compounds represent the main constituents, and account for that pleasant fresh fragrance filling the atmosphere of a pine forest.

Another strange-looking molecule is thujone, which contains the unusual and highly distinctive three-member ring fused with the familiar skeleton of menthone. Thujone is the main aromatic compound of sage: smelling good yet also poisonous, although not at the low doses used to add flavour to meats and other dishes.

From the garden filled with floral-smelling terpenes, we have moved slowly to the central market, where we are being regaled by the fresh odours of herbs, the menthol and piperitone of mint, camphor of rosemary, and thujone of sage. Before leaving the herb stall, let us take another look at the molecule of menthol. Instead of being swollen into almost spherical structures to get camphor odours, we can try to squeeze the six carbon atoms of the ring, forcing them flat, all on the same plane in what we call in chemistry an *aromatic* compound. These chemicals, which we have met earlier in this chapter, are derivatives of benzene or other structurally related rings, all flat and fulfilling certain structural requirements. But certainly this term comprises much more in its origins. In fact chemicals responsible for the aroma of spices, roasted meat, fried potatoes, and most cooked foods are aromatic compounds in the chemical sense, because they are similar to benzene, but the origin of the term refers to the pleasant smell of many of such substances.

A flattened menthol becomes thymol, and we can consider this molecule as the link between herbs (terpene derivatives) and spices (aromatic molecules). Its name obviously comes from thyme, for whose flavour it is almost solely responsible.

Thymol belongs to the class of phenols, compounds bearing a hydroxyl group directly attached to a benzene ring. The chemical properties of this –OH are quite different from those of an alcohol group as in menthol. For example, phenols are more acidic than alcohols, dissolve better in water, and react easily with a variety of chemicals. In particular, they are ready to undergo oxidation and several derivatives, both natural and synthetic, are used as radical scavengers. Many types of foods are rich in phenols.

Often their molecules are quite large and bear more than one phenolic group. Common examples are the so-called 'tocopherols', also known as vitamin E, abundant in olives and olive oil, and a variety of compounds responsible for the bright colours of fruits and vegetables. These are called anthocyanes (from *anthos*, Greek for flower) and occur in grapes, red oranges, berries, and several other foods. The common saying that brightly coloured vegetables are good for your health contains more than a grain of truth, as these phenolic compounds reduce the action of free radicals, responsible for the ageing process.

Phenols are also endowed with disinfectant properties. Phenol itself (a benzene ring with the hydroxy group and nothing else: Figure 2) is a common disinfectant, together with kresol (bearing in addition one methyl group attached to the ring) and other similar derivatives (Figure 7). The smell of kresol can be appropriately defined as the smell of disinfectant and had become a typical characteristic of the traditional London red telephone boxes (now fast disappearing), where its strong odour often blended with that of stale urine, created a distinctive bouquet.

Another similar chemical is guaiacol, chief component of cough syrup and giving this medicine its familiar smell. Some phenolic derivatives find their way into cosmetics, which they improve with

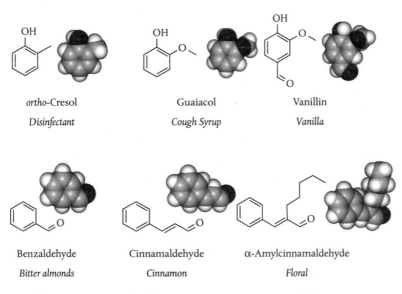

ortho-Cresol

Disinfectant

Guaiacol

Cough Syrup

Vanillin

Vanilla

Benzaldehyde

Bitter almonds

Cinnamaldehyde

Cinnamon

α-Amylcinnamaldehyde

Floral

Figure 7. From the pungent phenol we can ideally build o-cresol and guaiacol, remainders of disinfectants, by introducing a methyl or a methoxy group. Further addition of an aldehyde group generates the sweet pleasant aroma of vanillin. The structure of anethol, the captivating odorant of star anise, also falls into this group of molecules. Other aromatic aldehydes are present in spices and are endowed with interesting notes, from benzaldehyde, the typical smell of bitter almonds to cinnamaldehyde, the unique component of cinnamon scent. It is rather surprising to observe that the addition of a five-carbon chain to cinnamaldehyde profoundly modifies its odour which becomes floral.

their antioxidant and bacteriostatic properties, while contributing to a pleasant odour. In fact, while phenol and kresol are not too attractive owing to their pungent smell, other members of this family, representing additional groups on the benzene ring, present well-defined, appealing notes.

Eugenol and vanillin are two familiar examples: the first, already described (Figure 4), is the powerful odour of cloves, dominating the spice stalls of markets; the second, obviously, presents the pleasant sweet smell of vanilla. Both volatile compounds are the natural substances responsible for the characteristic notes of cloves and vanilla pods (Figure 7).

These two chemicals and similar derivatives can be found among the breakdown product of lignin, which, together with cellulose, make the solid structure of wood. We have already observed how it is actually from the wood of barrels that eugenol and other phenols are released into wine during storage, conferring that typical oak quality to aged wine.

Vanillin can be obtained from an unlikely source: a few years ago, the Japanese scientist Mayu Yamamoto presented a way of obtaining vanillin from cow dung. In fact the excrement of herbivores contains large amounts of lignin, which can be easily transformed into vanillin. Yamamoto won an Ig Nobel Prize for her very original contribution and was quick to note that the product extracted from faeces, which was after all the same molecule present in vanilla beans, could be used in products such as shampoo and aromatic candles, but perhaps not in food. Nevertheless, an ice cream maker in Cambridge Massachusetts introduced a new vanilla flavour, named after the Japanese scientist, without revealing the origin of the ingredients.

Remaining among the spices, we find anethole, the powerful scent of star anise, another spice much appreciated in cooking. And then we are attracted by another familiar absorbing smell, that of cinnamon. The structure of cinnamaldehyde, the chemical responsible for this flavour, is in some way related to benzaldehyde, the simple compound smelling of almonds. Just insert two carbon atoms connected by a double bond between the benzene ring and the aldehyde group, and the odour turns from almond to cinnamon, clearly different, but in some way related.

It is curious that if we add a chain of five carbon atoms to the molecule of cinnamaldehyde, attached to the carbon next to the aldehyde group, the typical spicy note is completely lost, leaving room for a most delicate fragrance. The scent of α-amylcinnamaldehyde is definitely floral and is much appreciated in perfumery (Figure 7).

When we compare the chemical structures of these spices, we can immediately perceive that they are all related to one another, a benzene ring as the common core equipped with two or three groups chosen from a limited number of possibilities. As in the case of floral fragrances, the smells of the spices are all different, but they are certainly related to one another.

Let us now leave the spice stall and, without going out of the market, follow the aroma of roast meat, equally pleasant and mouth-watering (to those of us who are not vegetarians). The molecules which make barbecues so attractive are still aromatic compounds (in the chemical sense), but the core is the ring of pyrazine, a modified benzene, where two carbon atoms at opposite positions in the ring are replaced by nitrogens. There is a large variety of pyrazine derivatives, sustaining small groups on the ring in different arrangements. We can find methyl and ethyl groups (one or two carbon atoms equipped with the necessary hydrogens), an acetyl group, and a methoxyl, mixed in any possible combination.

All these compounds are endowed with pleasant notes that are commonly indicated as *roast, fried, toasted*, but can be specifically recognized as *fried potatoes, roast peanuts, popcorn, fresh bread*, and many others. Roast coffee is very rich in such compounds and its volatile element, responsible for the pleasant aroma, contains hundreds of different chemicals, each contributing in some way to the overall impact.

Not all these aromatic compounds are pyrazine derivatives. At least two other rings, as flat as benzene, contribute pleasant-smelling molecules to coffee, as well as to cooked meat and other foods. These are rings of five atoms. One is called furan and is made up of four carbons and an oxygen; the other is thiazole and contains three carbons, one nitrogen and one atom of sulfur. We met sulfur before in connection with foul-smelling chemicals, but in thiazoles this atom demonstrates a completely different aspect, absolutely devoid of bad notes.

Both pyrazine and thiazole derivatives are produced in foods during cooking and originate from protein and carbohydrate breakdown products. Furans, instead, containing only carbon and oxygen (apart from the always present hydrogen), are the cyclization products of sugars. Representative examples of pyrazines, thiazoles and furans produced during cooking of different foods are shown in Figure 8. They all contain short chains of carbon atoms linked to the aromatic rings. In all three systems, if we increase the length of one of the chains we end up with a marked shift in the aroma, from roast, nutty, and caramel to green, fruity, and minty odours.

All of these compounds, which are attractive and mouth-watering, send us clear and important messages. First, they advertise the presence of important components in foods, proteins, and carbohydrates. Then, being synthesized only at high temperatures, they make cooked food more appealing. Probably it was because of such pleasant aromas that we adopted the practice of cooking many foods, making them at the same time healthier and easier to digest. In fact, cooking destroys potentially harmful micro-organisms, inactivates poisonous or anti-nutritional compounds present in foods and makes proteins more exposed to degradation, having been denatured, which means loosened and unfolded, by heat treatments. Of course high temperatures may sometimes have adverse effects on some vitamins, particularly vitamin C and other nutrients. For this reason, we generally prefer eating fruit and some vegetables raw. Their natural flavour is good enough and is often not improved by cooking.

If we move to larger molecules, which we can by travelling in different directions, we encounter further smells, some already familiar, others new. To provide a few examples, among molecules of twelve and thirteen carbon atoms we find geosmin, the earthy

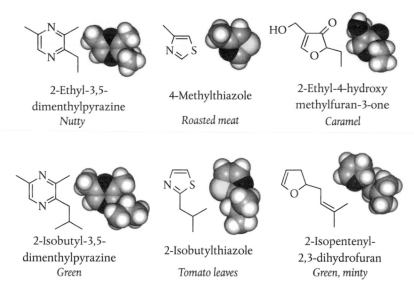

2-Ethyl-3,5-
dimenthylpyrazine
Nutty

4-Methylthiazole
Roasted meat

2-Ethyl-4-hydroxy
methylfuran-3-one
Caramel

2-Isobutyl-3,5-
dimenthylpyrazine
Green

2-Isobutylthiazole
Tomato leaves

2-Isopentenyl-
2,3-dihydrofuran
Green, minty

Figure 8. Examples of aromatic compounds common in our foods. Most of the volatile chemicals produced during cooking contain an aromatic ring, such as pyrazine, thiazole, and furane, bearing small groups attached in different positions. Their odour characters are similar and are described with terms like 'roast', 'nutty', 'caramel', 'toasted', 'burnt', and others. If we extend one of the carbon chains to between four and six atoms, we experience a marked change of odours towards 'green', 'floral', 'minty', 'fruity' qualities. We can also observe that, by strange coincidence, not uncommon in nature, some of these compounds are also produced by insects and used as pheromones: several methylpyrazines are alarm pheromones for some species of ants, while 2-isobutenyl-3-methylfuran is a pheromone for some acarid mites.

smelling compound introduced earlier (Figure 3), noting its similarity in odour and shape with the mushroom odorant 1-octen-3-ol. We also find β-ionone, a rather complex molecule smelling of violets, α- and β-santalol, the odoriferous compounds of sandalwood and others (Figure 9). The unique scent of this tree contains earthy notes together with an amber smell, another type of odour much valued in perfumery. A great deal of research was undertaken in the past to investigate these two very special scents, sandalwood and amber.

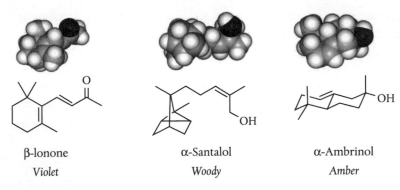

β-Ionone
Violet

α-Santalol
Woody

α-Ambrinol
Amber

Figure 9. Relatively large molecules with 13–16 carbon atoms exhibit pleasant odours, with floral, woody, and amber notes.

Both were originally found, as is the case of most types of odours, in nature. Ambergris is a secretion of sperm whales, and is occasionally found in huge lumps floating in the sea. The original compound present in this product, amber, is odourless, but its degradation products, ambrox and ambrinol, are among the most sought after odorants in perfumery. However, ambergris is very rare and cannot be regarded as a reliable source of products for the perfumery industry. Sandalwood is a very slow-growing tree and is becoming an endangered species because of its massive exploitation.

These facts prompted much research aimed not only at reproducing the natural molecules in the chemistry lab, but also at designing and synthesizing new chemicals endowed with similar olfactory notes. Such research also stimulated active interest in understanding how molecular structure was related to odour, while providing perfumers with new scents to use in their creations. In this way science and art cooperated in the exploration of our sense of smell both from chemical and psychological perspectives.

In our wanderings through the quarters of this city of smells, we are approaching the outskirts, the areas where we are likely to find the largest molecules still able to stimulate our nose. In fact, there is an upper limit to the size of odorants. Very large molecules are not volatile enough to float in the air and reach our nose. On the other hand, even if they could, they would not find receptors with cavities big enough to accept and recognize them. Clearly, there was no need to equip our sense of smell with detectors for molecules so heavy they could never get close to the nostrils.

The region of large molecules is dominated by only two types of odour, contrasting with each other: the alluring scent of musk and the repulsive stench of stale urine.

We were introduced to the molecules responsible for such smells in Chapter 1. Musky odorants include natural compounds, often the sex pheromones of some mammals, but also found in plants, as well as a variety of different chemical structures (Figure 10). We have already discussed how the need to protect endangered species and the high demand for this distinctive fragrance from the perfumery industry prompted and supported very active research aimed at understanding relationships between odour and molecular structure, and at the same

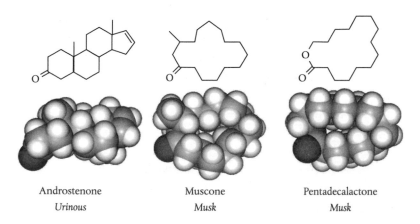

| Androstenone | Muscone | Pentadecalactone |
| Urinous | Musk | Musk |

Figure 10. The largest molecules able to stimulate our nose are the sources of two types of contrasting smells. Androstenone (19 carbon atoms) is a by-product of testosterone and has a strong and aggressive urinous smell, while the other two large cyclic structures are endowed with extremely pleasant musk notes, highly appreciated in perfumery.

time designing and synthesizing new chemicals mimicking the odour of the natural substances.

Musk odorants include very different structures, from macrocyclic ketones and lactones to tricyclic compounds and aromatic nitroderivatives. Such wide flexibility could be accounted for by the large size of these odorants and the correspondingly large size of the specific olfactory receptor. This means that a new molecule of the size of a musk odorant either fits into the musk receptor or cannot find any other receptor to stimulate, consequently being odourless.

There is however a single exception: the urinous receptor. Recall the structure of androstenone, that powerful smelling steroid, which originates from testosterone, the male hormone, and is present in the urine (Figure 1). We also noted that this same compound is a potent aphrodisiac for the sow, being released into the saliva of mature boars. There are not many synthetic compounds mimicking this repulsive smell, most likely because of the lack of interest in its very particular olfactory note. However, our rejection of this odour is certainly a product of culture, since it is associated with urine, which we associate with a dirty environment. But for many mammals, urine is used in communication between sexes and the presence of androstenone clearly indicates the presence of a male in the area.

We are at the end of our explorative tour and it would be a good idea to climb a hill overlooking the city to get a bird's eye view of the areas we have just visited. There is the market, right in the centre with its spice vendors, barbecue stalls, the woman selling fried potatoes and popcorn...all flat molecules, aromatic compounds, derivatives of benzene, pyrazine, thiazole, furan, and a few others. A few short stubs protrude from the core of these molecules; together they contain no more than 10–12 major atoms (including carbon, nitrogen, oxygen, and sulfur). At the edge of the market we find herb shops, with some aromatic compounds, like thymol, but most molecules are not flat: there are curved rings looking like chairs and bearing some side chains, like menthol. These notes become turpentine, camphor, and balsamic as we enter the park nearby with its large trees. All these molecules contain the 10 carbon atom basic structure of terpenoids, but appear more round in shape. From the herb stalls, if we turn our eyes to the other side we find a garden full of floral scents, geraniol, linalool, citronellol of rose and geranium mingling with the citrus notes of citral and citronellal of lemon trees and orange blossoms. From these open-chain structures we can glance at other linear and branched alcohols, aldehydes, or acids in three parallel streets, each exhibiting chemicals of the same type, arranged according to the length of the chain. Further out from the city centre the odours of larger and more complex molecules dominate—woody musk (we can imagine some exotic animal), and the stench of stale urine from the lavatory of a long disused railway station.

In this imaginary trip we have glimpsed some basic relationships between different types of smell and the molecules behind them. But we are very far from establishing useful and informative correlations, which would enable us to predict the smell of a chemical compound just by looking at its molecular structure. Many are the variables and subtle are the elements making each smell unique. But we should not be disappointed: it is such complexity of the olfactory experience and the extreme difficulty of classifying smells into an organized frame-work which makes the creation of perfumes or the cooking of foods an art more than a science.

THE OLFACTORY CODE

A Chemical Language

THE LANGUAGE OF SMELL

In Chapter 2 we met several types of odorant molecules and became acquainted with their diverse characters and distinctive odours. Chemists have the habit of smelling everything, poking their noses into test tubes which contain a novel compound, as a first and immediate way of being introduced to a new character in the chemical world. In older chemistry papers, but also often in more recent ones, when a new compound is synthesized its odour is also recorded, together with other characteristics, such as the structure, the boiling or melting point, and a variety of spectra.

But, while the colour or other properties of a compound can be deduced from its chemical structure, the odour represents new information. We have already stressed this point, showing how we cannot talk about odour without a direct reference to the nose, specifically our nose. In other words, the odour appears only when volatile molecules physically meet specific proteins inside our olfactory system. This means that the odour of a molecule cannot be predicted just by looking at its chemical structure. In other words, we cannot imagine what a molecule smells like if our only information is the chemical structure.

But, with experience we can learn how to relate smell to molecular structure and build a database of information, which in the end might enable us to guess the odour of a new molecule. The major problem is the extreme complexity of the olfactory code and, strictly related to this difficulty, is our ignorance of how we distinguish smells from one another. Biochemistry and molecular biology are currently trying to lift this veil of ignorance and mystery which, until very recently, surrounded our sense of smell, but the road is long and we are just at the beginning.

Nevertheless, we can start to learn the olfactory meaning of molecules, in the same way as we learn to read letters written in an alphabet or understand the meaning of ideograms. We can master this new language written with molecules, but it is a very difficult language because rules are scarce and neither clear nor general.

Anyone who starts learning Chinese, even native Chinese people, have to confront the difficult task of learning two languages at the same time, the spoken words and the written characters. Unlike an alphabetical language, there is not much relationship between sounds and ideograms. We need to associate the correct sound and its meaning to each character and only after a long and painstaking application will we be able to *read*. During such a process we become aware of the fact that some relationships do exist and sometimes we can guess the pronunciation of an ideogram, but only to a limited extent.

With smells it is not so different. There are more than 40,000 Chinese ideograms, but with only 10 per cent of them we can read books and newspapers. There are countless molecules, but they can be reduced to a reasonable, although still very large, number when we group them into classes and sub-classes. To learn the language of smells we have to try at first to associate different odour characters with the appropriate molecules. This becomes almost instinctive for chemists, particularly those working in the field of perfumery or food flavours, but it is no easy task. Even perfumers and experts in food flavours, although good at distinguishing subtle differences in the scent of flowers or in the bouquet of a wine, have great difficulty when they try to guess the molecules at the origin of those sensations.

DECIPHERING THE CHEMICAL LANGUAGE OF OLFACTION

Complexity of olfactory messages

In fact, olfaction reminds us of a complete language, made of sentences, where several words are combined to create infinite possibilities of expression, not excluding the addition of neologisms or words imported from other tongues. For instance, new synthetic molecules with previously unexperienced odours can be regarded as the neologisms of the olfactory language. There is also the equivalent of grammar and rules of syntax, with chemicals producing different perceptual or behavioural responses according to the context. In fact, our emotional reaction to some smells may depend on the situation in which we perceive them. We have already reported the case of isovaleric acid (and other fatty acids), whose odour is repulsive when its source is an individual who has limited familiarity with soap and water, but can be highly appreciated if part of the aroma of some special cheeses.

And what about dialects and idioms, words and expressions used with slightly different meanings by different communities? Well, each of us detects smells in a different way. We have come across the phenomenon of specific anosmia in the first chapter, a characteristic extremely common and varied among human beings, which affects the way we perceive smells. In a more general way, our different sensitivities to each of the many basic odours produces in the end olfactory images which are different for each of us and are responsible for unique sensations.

Thus, we can think of the bouquet of a wine as a poem, where words (the volatile molecules) are combined and presented in a special fashion, and each word in turn is the harmonious product of letters of an alphabet (the elementary interactions with specific receptors), chosen and arranged to convey desired meanings. From such a perspective, we should not be surprised at our different personal preferences for specific wines or generally for the foods we eat.

If such elements of a complex language can be found in human olfaction, in most other animal species, which have not developed a communication based on sounds, chemical information takes the place of a real idiom which can reach in some cases a high degree of complexity. This is particularly true in social species, such as honey bees, ants, and termites, which live in large communities organized in strictly regulated ranks and hierarchies with specific tasks assigned to each member. In such societies it is not only recognition of different castes that is mediated by olfactory signals, but also the duties and commands which regulate the life of the community, as well as information about food, the needs of larvae, the presence of foreign individuals and other dangers, are all conveyed by specific chemical messages. We are therefore dealing with molecules that take the place of words, each bearing a specific meaning and each part of a complex language.

Therefore, as in a spoken language, we can expect that sometimes the same word of this chemical alphabet might convey different meanings according to the context. However incredible it might sound in what appear to be relatively primitive animals, this is actually the case, and several examples have been reported where the same pheromone can elicit different behavioural responses depending on the presence of other olfactory stimuli or else of additional information coming through other sensory modalities, such as vision or hearing.

Breaking the olfactory code

Now, if we want to go to the core of olfaction and unveil its most hidden secrets, the key for us to understand why linalool smells of flowers and 3-octenol smells of mushrooms, we have to decipher the olfactory code, we have to identify the alphabet of this chemical language.

Breaking the olfactory code is a problem similar in a way to that of decoding an ancient manuscript written in an unknown language. When trying to decipher a mysterious text, we are faced with two types of problem: the language and the written symbols. Often we already know one of these two elements. For instance, Etruscan is written in the Greek alphabet, but the language was unknown until recently. Conversely, an encrypted message is likely to be written in English, but we have to decipher the meaning of each symbol. The worst and most hopeless case is when we do not know the type of language (the grammar rules and the meaning of the words) nor understand the function of each character: is it an alphabetic, a syllabic or hieroglyphic language? This was the case of Linear B until the 1950s and still is the case for Linear A, two of the languages of ancient Crete.

When confronted with an unknown text, the first question is: what sort of language are we dealing with: is it alphabetic, like English and most of our languages, or syllabic like Japanese, or else based on ideograms, like Chinese? The key which allowed Champollion to decipher Egyptian hieroglyphs was the intuition that, contrary to the current belief, the major part of the hieroglyphs were not pictograms, but elements of an alphabet. In other words, most Egyptian hieroglyphs represented sounds rather than concepts. From this point on, it was relatively easy to recognize first the names of the people cited in the text and then one by one all the elements of the alphabet.

Using similar logic pathways, chemists, who were the first to undertake the complex task of decoding the language of odours, have formulated hypotheses and verified them against olfactory experiences, in order to break the olfactory code, like deciphering a chemical Rosetta stone, by establishing relationships between familiar and well-defined odour types with the structural parameters of odorants.

The first question to ask is: what are the molecular parameters that our nose—or any other olfactory system—considers to be relevant for decoding an odour message. We can list physical properties (melting point, boiling point, refractive index, solubility, colour) and chemical properties (acidity, susceptibility to oxidation or reduction, specific reactivity towards certain substances), as well as structural elements (size, shape, nature, and position of functional groups), to describe a molecule. But, out of all these aspects, which are relevant to the nose, which ones are recognized by olfactory receptors and can therefore be utilized to predict the smell of the molecule? Certainly not all of them. Once we have recognized which molecular properties can best be related to smell, we are on our way to understanding the type of language used in chemical communication.

Sharing olfactory experiences

The first tool we need is a system for classifying smells, in order to group them, and establish distances and relationships. But it quickly becomes apparent how difficult it is to communicate our olfactory experiences. To classify smells we need words, descriptors that enable us to exchange and compare our sensations. These can be simple words, taken from our daily experience, words which have always been used to indicate common objects. We employ expressions such as the scent of apples, of jasmine, of nuts, roast meat, damp soil, freshly baked bread, vanilla, and many others. These terms do not try to describe our subjective olfactory perceptions, but they simply refer to the sources of smells.

This is not surprising, as we also resort to such stratagems in other situations, for example when we want to describe colours. We talk about sky blue, leaf green, lemon yellow, blood red, brick red. In these cases, to be more specific, we refer to a familiar object endowed with that particular shade of colour.

No wonder, then, that we use names of common objects—well known to everybody—to define smells, which comprise a variety and diversity much wider than colours. Therefore, although smells are carried by molecules, we never think of referring to the flavour of banana as amyl acetate or to that of mushrooms as 1-octen-3-ol. This way of communicating our olfactory experiences by referring to the object which caused the experience is quite crude and inaccurate. In fact, each of us perceives the same chemical or blend of chemicals in unique, although similar ways. In music, the written notes can be played and interpreted in different ways by the performer or by the conductor.

A suitable linguistic comparison would be the ancient texts used by the Nashi (also called Dongba) minority people, who inhabit the northern region of Yunan in China. The ancestors of present-day Dongba used pictograms, in order not to write words but to remind the reader of concepts. In fact, around the mid-1990s a Dongba-Chinese dictionary was published, containing an explanation of each pictogram rather than a faithful translation. Our compilations of smell descriptions, such as the classic Arctander,[7] which associate an odour character to each chemical compound, can be regarded as that kind of dictionary.

Defining smells and giving names to our olfactory experiences is an important step towards understanding relationships between chemical structures and odour. For instance, sometimes we come across smells that are perceived as alike and immediately look for similarities in the structures of the corresponding molecules. In fact, the first approach to putting olfaction into a scientific frame was an effort to group odour types into families and to try to understand what was common to the members of each family in terms of molecular features.

Such an approach has led chemists to generate classifications based on a certain number and type of basic smells, called *primary odours* by analogy with the three primary colours of our visual system.[8] This was the main trend in olfactory research during the 1960s and 1970s, before applying biochemical approaches to the study of olfaction.

Based on everyday observations, several groups of scientists proposed their own list of primary odours. Unfortunately, there was little agreement even on the number of these basic elements, which ranged between as few as seven to more than 50. Even the largest number of suggested primary odours was still far away from the several hundred types of olfactory receptors discovered later by molecular biology.

In contrast to the forbidding complexity of the olfactory code, our sense of taste is incredibly simple and based on just five basic sensations. It was thanks to this simplicity that the elements forming the taste code had been recognized in ancient times, as was the case with the colour code before any scientific investigation. In fact, the classification of tastes into four categories—sweet, bitter, salty, and acid— had remained unchanged for centuries and has been recently confirmed by the results of molecular biology. A fifth basic taste, named with a Japanese term *umami* has been added in recent times and the relative specific receptor has been identified. It is the typical and familiar taste of meat broth, previously believed to be the effect of the combined stimulation of salty and sweet receptors. The stimuli for this taste modality are some amino acids, chiefly glutamic acid. It is interesting to observe that, even before this was recognized as a fifth taste, for centuries it had been common practice in oriental cuisine to add glutamate to make soups and dishes more appetizing. In western countries meat broth concentrates are used instead, which contain glutamate as one of the main ingredients.

So, simplicity is the common feature of the taste code as well as the colour code, and it was such simplicity that allowed such codes to be deciphered through an empirical approach, before any scientific investigation. And it certainly was this simplicity that misled scientists into thinking that olfaction also might be based on a similarly simple code.

But the difficulty experienced in classifying smells and decoding their chemical language should have been a clue that the olfactory code had to be very complex and based on a very large number of elementary sensations. It is also true that the idea of having hundreds of olfactory receptors in our nose was rejected by scientists for a long time as unlikely and not *economical*. After all, the sense of smell is less important than vision (at least for humans), so why would nature invest a lot of energy and synthesize a great number of proteins to detect odours, while only three receptors are enough to distinguish colours? Taste, on the other hand, can perform a similar task, that of analysing the chemical environment, with a limited number of sensors: a simple and economical solution, and also aesthetically more appealing.

Our aesthetic sense often strongly affects our vision of nature and prevents us from analysing the experimental data with the necessary objectivity. On the other hand, some observations seem to support the idea that nature is simple and beautiful. The DNA double helix is a typical example of golden simplicity and beauty. Its most important property, the ability to duplicate itself, is strictly related to its essential and robust architecture, as harmonious and solid as a romanesque church. If we wanted to design a self replicating molecule, we would hardly end up with anything simpler than the four bases of DNA.

Biology, however, can be extremely complex behind a facade of beauty and simplicity. In some cases complexity and redundancy are not really functional, but are just the intermediate products of evolutionary processes, which have not yet reached the best solutions. In other cases, however, complexity is the only way to cope with a variety of biological aspects, although our incomplete knowledge still prevents us from appreciating the underlying logic. This is the case with olfaction and explains why it took so long for us to unravel the biochemical mechanisms of a phenomenon so important in our everyday experience.

How many basic odours?

As we have seen, the tendency of scientists to simplify, group, and build models as fundamental as possible, led to classifications of odours into no more than a couple of dozen elementary sensations, in contrast with the many hundreds of olfactory receptors utilized by mammals.

At this point, after the tools of molecular biology have finally identified the genes encoding olfactory receptors and revealed their unexpectedly large number, we should ask ourselves why we need such a complex and apparently redundant system. The answer was always in front of our nose and a better question would be why we did not find it earlier.

We have compared the olfactory code to the colour code, but let us take into consideration another sensory system, hearing. Our auditory organ contains several thousand different sensing elements, each tuned in to a specific wavelength. Why do we need so many sensors? Imagine we are in a crowded room, talking to a friend among many other people each talking and making noise, and perhaps there is also music in the background. We can still pick up the elements of our conversation and make sense of the sounds coming from the mouth of our friend, despite the very intense and diverse background noise. This is possible because we are endowed with such a complex hearing system.

If we only had a few sound receptors, responding to broad areas of the spectrum, as in the case for colours, then we would only be able to detect some averaged signals, in which all the sounds would be mixed without any possibility of discriminating the original components. It is essential, rather, that we are able to pick up each sound, each individual word, isolating them from the background, in order to carry on our conversation. The same complex system enables us to appreciate the sound of a violin in an orchestra or detect someone calling us from within a noisy crowd.

We do not need such a high-performance system for detecting colours. The basic difference between colour vision and hearing is that colours mix and we get averaged information, while sounds do not mix and we perceive all the individual original signals.

We can apply more or less these same considerations to olfaction and chemical communication. The environment is flooded with different odours and it is of vital importance to be able to detect and recognize a specific olfactory message in the middle of thousands of other chemical stimuli. Not so relevant for humans, it is true, but for almost every other animal species survival depends on the correct functioning of the olfactory system. It is of vital importance for a predator to detect the presence of prey by its odour, and even more for the prey to be aware of the predator and keep at a distance. A male insect flying around must be able to find the female of its own species sitting on a tree and advertising her presence by releasing specific pheromones into the environment. This is not an easy task amongst all the environmental odours, including pheromones of other insects of similar species.

Although for us humans the sense of smell is no longer essential for survival, the situation was probably different at the beginning of our civilization and to a certain extent until relatively recent times. It is true that we now mostly rely on the expiry date printed on our food packages to know if the content can still be safely consumed, but certainly our sense of smell can perform a chemical analysis on the spot and in real time on what we are eating and send us clear warning signals if our food is even slightly contaminated or there is an incipient degradation process.

We still use our sense of smell to detect a gas leak or a cake burning in the oven. And, of course, we can fully appreciate and enjoy our food or a glass of wine because we can detect all the subtle shades of aroma, combining to give us a rich and complex sensation rather than disappearing into a flat, grey taste mixture. Very appropriately, when describing the aroma of a wine we often refer to a symphony of olfactory and gustatory notes. All this would not be possible with only a handful of olfactory receptors. We can justifiably say, therefore, that a very complex olfactory system is essential for the survival of most animal species, while fortuitously helping us humans to appreciate the pleasures of life.

All the above considerations indicate that the auditory system would be a better model for olfaction, rather than the simpler and more aesthetically attractive colour code. We shall see, however, that the olfactory language, although being much more complex than the colour code, is not so specific and detailed as the recognition of sounds. We can conclude, therefore, that olfaction uses a strategy somewhere between that used by our auditory system and the one which discriminates colours, yet much closer to the former.

THE PSYCHOPHYSICAL METHOD

The long quest to crack the olfactory code and identify the basic elements of the chemical language perceived by our noses began with very simple observations and crude correlations. Thousands of new molecules have been synthesized in the course of several decades with the single purpose of studying their olfactory characteristics. Specific modifications were introduced in the structures of known odorants to monitor their effect on the perceived smell.

These studies certainly failed to decipher the language of odours, which still remains mysterious to a large extent, but they have produced a lot of interesting chemicals endowed with the desired scent and at the same time easier to make, more stable, safer, and cheaper than the natural compounds they are meant to mimic. Similarly, the combined effort of generations of chemists in their search among molecular structures and smells has contributed to the building of a very large database, which subsequently created solid foundations for the study of biochemistry and the molecular biology of olfaction.

The psychophysical method which was applied at the beginning was very crude and tried to build a network of correlations between molecular structure and odour, bypassing all the biochemical reactions and physiological events which translate chemical information encoded in the molecules of odorants into perceived olfactory sensations.

We can imagine the olfactory system, in its totality from nose to brain, as a black box, tightly closed and impenetrable, provided only with an entry port and an output. This mysterious box had not yet been cracked before the application of molecular biology began to work, and this allowed a lot of different stories to be conjured up, some with a grain of scientific evidence, others mostly the fruits of the imagination. At that time theories were debated rather than experimental data, and it seemed as if everybody was afraid of opening the sacred box, for fear of discovering a corpse or violating a magic spell.

Having been in the field long enough, I had the opportunity of witnessing the development of this research from the very beginning and to feel the excitement of discovering even minor hints which might suggest what was inside that mysterious closed container. Around the early 1970s biochemistry was already an established science and receptors for neurotransmitters and hormones were being discovered and identified. However, experimental methods were rather crude and required relatively large amounts of biological material. Molecular biology was at its inception and the era of genomes was not even in our dreams.

Therefore, rather than going to the core of the problem and studying the proteins involved in the recognition of odours—perhaps at that time applying biochemical methods to the study of olfaction seemed too risky and difficult—scientists, for the most part organic chemists, chose to investigate the mechanism of odour coding without opening the box, but by interrogating the system with appropriate questions.

The questions were presented in the form of volatile molecules which each reached the nose with a different message. The answers were then collected from human subjects as verbal descriptions, using common words and expressions to communicate their olfactory experiences. Comparing the answers with the structural elements of the odorants we could get information on the type of code used by the nose to read chemical messages.

We can retrace these first steps in the light of what we have learned in more recent times with the help of molecular biology, which provided us with the right tools to open the black box and directly observe what was going on inside.

DECODING OLFACTORY MESSAGES

The key step in olfactory perception is decoding the chemical information carried by odorant molecules and translating it into electric signals to be sent to the brain. We now know that the peripheral olfactory neurons, in particular the receptor proteins sitting on the membrane of their protruding cilia, are responsible for this translation. An intuition of this kind provided the key that paved the way to understanding olfaction.

At the beginning of the 1960s, John Amoore compared hundreds of molecular structures and tried to establish correlations with the odours elicited by the compounds they represented.[9] His conclusion was that the molecular characteristics that best correlate with olfactory properties are the size and shape of molecules, in other words, structural parameters.

This ingenious idea represented a landmark in the scientific studies of olfaction. From then on, the attention of chemists working in the field of odours was focused on structural characteristics of molecules. It finally became evident where we should look for correlations between smell and molecular properties. The strategy for designing new molecules with a desired smell was also better focused and the path to follow more clearly indicated.

In some ways, these strategies tried to apply to the study of olfaction techniques already established and validated for the design of new drugs, known as the *pharmacological approach*.

In fact, olfaction is just another example of chemical communication, using mechanisms similar to those regulating all types of interactions between small molecules and proteins, from enzymes to receptors and other classes of binding proteins.

The pharmacological approach

Let us first take a brief look at the strategy that has proved very successful in the discovery of new drugs, greatly contributing to the advancement of knowledge and at the same time providing a large number of new chemicals for the treatment of mental and hormonal disorders, as well as other diseases.

The development of a new drug usually starts with the structure of a known compound endowed with an established physiological effect and aims at the design of novel chemicals exhibiting the same physiological effect and at the same time different physico-chemical characteristics, by introducing small structural variations in the molecule and recording their contribution to the final physiological outcome.

Some examples would clarify this method, but let us first get familiar with a few basic facts of biology. We know that individuals of the same species use chemical messages to communicate with each other. Insects and mammals use pheromones to advertise their presence to their partners, to warn of a specific danger or of the existence of food, while social species, like honey bees or ants, have developed a rich and complex system of communication mediated by molecules.

Similarly, the cells of our bodies utilize a chemical language to exchange information and regulate their activities. Our brain, for instance, is made up of special cells, the neurons, connected to each other through a very large number of tentacles, called dendrites. Each neuron communicates with many others through electric signals and by means of a very large and complex network. However, when we look deeper, we find that this is also a type of chemical communication. In fact, there is no direct physical connection between neurons. Between the tip of the dendrite sending the signal and the receiving neuron there is a little gap, across which the transmitter neuron releases certain chemical compounds—the neurotransmitters.

These soluble molecules travel to the target neuron, where receptors sitting on the membrane identify each molecule and send specific messages inside the cell, in which a chain of biochemical events is activated, leading eventually to an electric signal. The generated electric signal, in turn, stimulates the secretion of other neurotransmitters, which will excite yet other neurons, and so on. In practice, we are dealing with a system that recognizes the information encoded in chemical structures and translates such messages into electric impulses.

The olfactory system also works, to some extent, in a similar way. It monitors the environment, catching the volatile molecules carried by the air, identifying each of them and sending appropriate messages to the brain. There are many other systems of chemical communication active in our body—such as those mediated by several hormones—

but we can hold on to the example of neurons and neurotransmitters to illustrate the pharmacological approach and verify to what extent we can apply such a method to the study of olfaction.

In practice, we stimulate a single type of receptor with a range of synthetic drugs, designed on the structure of the endogenous active compound and compare the responses obtained with that produced by the natural chemical. Although we can study receptors *in vitro* after isolation and purification, the pharmacological approach can also be applied to a live organ or organism. In such cases, we stimulate the whole system with a range of structurally related chemicals and observe a certain physiological response, such as the contraction of a muscle, or a change to the heart rate.

Between the stimulus and the response there is a complex organism and a complex chain of biochemical events; however, we know that our drugs can act only on a particular target, the specific receptor we are investigating. This fact allows us to evaluate the efficacy of each drug just by monitoring the intensity of the response, because all other receptor systems present in our preparation do not interfere with our measurements.

Structure–odour relationships

All this appears rather simple and straightforward, but the question is: can we apply the same method to olfaction? Is it possible to draw simple relationships between structural parameters of the odorant molecules and their perceived smells? We have already discussed the extreme complexity of the olfactory language, how mixtures of odorants can elicit novel sensations in the brain, just as combinations of words produce new meanings and concepts. We have also observed how the same chemical can generate different olfactory sensations according to context.

All this refers to the processing of the peripheral signals. But what of the first recognition and translation of the original chemical messages? Again, we are faced with another level of complexity, which adds to what we have discussed earlier. This type of complexity can be put down to three main features of the interactions between odorant molecules and receptor proteins: (1) most odorant structures are flexible and can present different aspects to the receptors; (2) olfactory receptors are not too choosy, unlike those for neurotransmitters or hormones, and can accept several odorants of similar structure; and (3) we are equipped with hundreds of different receptors, each interacting, to different extents, with several potential volatile chemicals: this is what we call the *combinatorial code*, which generates very complex and diverse patterns of responses even when the stimulus is quite simple in its composition.

Therefore in olfaction a simple linear correlation between molecular structure and perceived odour cannot be expected. In fact, because each chemical substance can interact with several types of receptors at the same time, the response it produces is the result of many components. If we now introduce a modification in the molecular structure of an odorant, we might reduce its affinity for a specific receptor, but at the same time the new molecule could interact more strongly with other receptors. Therefore, the effect of such modification cannot be measured by the intensity of the odour perceived, because its quality is strongly affected. Measuring the quality of a smell, on the other hand, represents a very difficult task.

Finally, there is another problem that makes things more difficult when we try to apply the pharmacological approach to the study of olfaction. We have no information on which molecular structures best fit each odorant receptor. When designing a new drug, on the other hand, the reliable reference is the chemical structure of a neurotransmitter or a hormone or any other natural compound known to be responsible for the physiological effect we observe.

So we are confronted with a problem that is too difficult and our response to that is to try to simplify the terms of the questions, while at the same time building simple models that could describe our systems. This approach works in many cases and at least provides us with the basic information. When dealing with very complex mathematical equations to describe a biological system, for instance, the first thing we do is to introduce approximations, deleting terms giving minor contributions in order to produce a manageable formula. The model we obtain may not be very accurate, but it can at least reproduce the basic features of the system and make crude predictions, as a first step in our endeavour.

We can illustrate this procedure with some simple examples in which we modify specific elements in the structure of an odorant and observe how the odour has changed. Such an approach will reveal the molecular parameters most important for a certain odour and, at the same time, will provide detailed information for designing new molecules with a specific odour.

Let us start with some classical cases showing that the odour of a chemical does not change much when we replace the functional group with a different one, while keeping the skeleton of the molecule unaffected. Benzaldehyde, a simple compound made of an aldehyde group attached to a benzene ring, has a typical smell of bitter almond (Figure 11). If we replace the aldehyde function with a nitro group or with a nitrile we end up with chemicals different from benzaldehyde in their chemical properties, but very close in odour.

In a similar way we can replace the aldehyde group of citronellal, a citrus-smelling compound, with a nitrile and obtain citralva, endowed with virtually the same citrus note, but much more stable to oxidation and degradation (Figure 11).

We can also recall some examples of this kind illustrated in Chapter 3. While examining derivatives of three aromatic rings, pyrazine, thiazole, and furan (Figure 8), we observed how the smell can range from nutty and roasted to green and vegetable, depending on the length of the protruding carbon chains, while it is not much affected when we replace one aromatic ring with another.

One more example is provided by the series of γ-lactones (Figure 12), in which the length of the carbon chain has a profound effect on the odour. The first one with only a single carbon attached to the ring, smells like freshly baked bread, the second with a chain of five carbon atoms presents a typical note of coconut, while the odour of

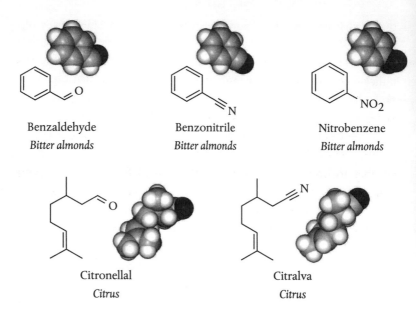

| Benzaldehyde | Benzonitrile | Nitrobenzene |
| *Bitter almonds* | *Bitter almonds* | *Bitter almonds* |

| Citronellal | Citralva |
| *Citrus* | *Citrus* |

Figure 11. Classic examples supporting the theory that odour mainly depends on stereochemical parameters. Replacing the aldehyde group of benzaldehyde with a nitrile or a nitro group does not affect the main olfactory note. Similarly, the citrus scent of citronellal is retained in the nitrile derivative citralva. In this case, as in similar ones, the replacing of the labile aldehyde function with a very stable nitrile group brings practical advantages.

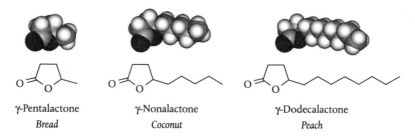

| γ-Pentalactone | γ-Nonalactone | γ-Dodecalactone |
| *Bread* | *Coconut* | *Peach* |

Figure 12. Chemical structures and molecular models of three γ-lactones with different odours. The three molecules contain the same ring, but differ in the length of the side chain. The first molecule bears a single carbon as side chain and is endowed with a note described as 'freshly baked bread'. When we increase the chain to five carbons, the odour becomes that of coconut, and a further increase produces peach and other fruity notes.

the third, bearing eight carbons in a row, is fruity, specifically smelling of peach.

These examples and many others available in the chemical literature indicate that the odour strongly depends on the size and shape (the skeleton) of the molecule, rather than the type of functional group. We can certainly exploit this property to design odorants endowed with the desired smell, but at the same time more stable than the natural chemicals, as is the case of citronellal and citralva, or easier and cheaper to prepare, or else with other useful features, like a better affinity for clothes (a highly desirable property in fragrances to be added to detergents) or for the skin (when used in perfumes), combined with safety for humans and for the environment.

The strategy would be to replace the functional group with a different one, as we have seen, but we could also modify the scaffolding of the molecule in some cases, while keeping a similar overall shape. This last concept can be illustrated by an interesting study I found myself involved in while collaborating with my chemist colleagues and friends Elio Napolitano at my own University of Pisa and Cecilia Anselmi at the University of Siena. We have come across geosmin (Figure 3) more than once, the earthy-smelling compound, unwanted in drinking water, but highly sought after in perfumery. Its odour is very attractive and extremely strong, but quite difficult and expensive to obtain.

At that time I was interested in geosmin also for more speculative reasons. In fact this molecule has been postulated as a marker for water. Of course water is extremely important for the survival of all living organisms, but we cannot smell water, because our olfactory neurons are continually immersed in water. Therefore, we can guess where we are likely to find water by smelling volatiles which are produced in association with its presence. Geosmin is a good candidate for being such an indicator, as it is typically produced and released after rain.

Geosmin was not commercially available at that time and the only way of obtaining a sample was either to purify it from micro-organisms or to invest time and energy on a long multistep synthesis, in both cases with extremely small yields. Therefore, we decided to explore another alternative and design new chemicals simpler in structure and synthesis, but with similar molecular shapes. As the main difficulty was to synthesize the two-ring system, we first designed structures containing only one ring which were much easier to synthesize. These can be ideally derived from the molecule of geosmin by *opening* one of the rings. Some of them retained the earthy smell of the natural compound, so we resolved to go a step further and try even simpler molecules containing no rings. The result was some new chemicals endowed with earthy smell and made in one-step syntheses starting from cheap commercial compounds (Figure 13).

But something was inevitably lost in this process. The pleasant earthy character of these molecules was accompanied by secondary weaker notes of camphor, which became noticeable when smelling the pure compounds or highly concentrated solutions. This effect was likely related to the increased flexibility of the new structures, which can consequently adapt to and stimulate receptors for the camphor smell. We know that this olfactory character is generally associated with molecules of about 10–12 carbon and round shape.

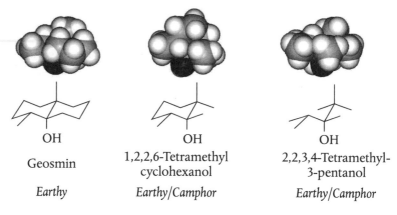

Geosmin	1,2,2,6-Tetramethyl cyclohexanol	2,2,3,4-Tetramethyl-3-pentanol
OH	OH	OH
Earthy	*Earthy/Camphor*	*Earthy/Camphor*

Figure 13. An example showing how the information on structure–odour relationships can be applied to the design of new odorants. The cyclohexanol derivative and the open-chain tertiary alcohol still retain the earthy note of geosmin, while being much easier and cheaper to synthesize.

Such types of correlations are relatively easy to establish and can be verified through the design and synthesis of new molecules endowed with a target odour. The approach has been widely exploited in the field of perfumery, as new chemicals can be relatively easily accepted in the formulation of a perfume or a detergent, while their uses as food flavourers raises serious concerns and the regulation is much stricter.

The freedom to build new molecules endowed with novel odours produced invaluable tools for a new form of art, where the creativity of chemistry and the unpredictability of odours has provided boundless fields to explore in the world of chemical structures.

PART 2

MESSENGERS OF SEX AND DANGER

5

INSECT PHEROMONES

Fatal Attraction

LOVE, WARNING, DECEPTION

The magic of chemical attraction

If you visit a forest in North America or stroll through a pine wood along the Italian coast or just walk along a tree-flanked road in many parts of the world, carrying with you a special chemical known by its commercial name as *disparlure*, it is likely that you will be quickly surrounded by a number of moths nervously fluttering about your head.

The insects are all males of the species known as gypsy moth or *Lymantria dispar*, which is a major pest and has been responsible for destroying entire forests. The chemical you need to perform this piece of magic is the species-specific sex pheromone. This compound is synthesized by females in special glands and released in the environment. It is an extremely potent aphrodisiac and you only need what is left on your fingers after touching the exterior of the vial to attract males from very far away.

Several years ago, when I started my research on the proteins of the olfactory system in insects, to collect the gypsy moths I needed for my experiments I would just walk along the river bank lined with holm oaks (*Quercus ilex*), carrying a jar with an invisible trace of this phero-mone. The males would come from every direction and, after a brief search would dive into the jar and stay there rapidly flapping their wings. Old couples strolling along and mothers pushing their prams would stop and stare, incredulous at what they were witnessing.

Of course there was no magic. The moths were just following the scent of their females. What is unique is the extreme sensitivity of insects to their pheromones. Among the woods of southern Germany, not far from Munich, in the Max-Planck Institute for Behavioural Physiology, already famous for having hosted Konrad Lorenz and his geese for many years, Karl-Ernst Kaissling performed pioneering studies on insect pheromones for many years.[10] He applied electro-physiological techniques to record the responses of single olfactory sensilla on the insects' antennae when challenged with a pheromone or any other volatile compounds. By lowering the concentration of the smell until an electrical signal could be detected, he calculated that as few as 10 molecules of the specific pheromone, bombykol, are enough to stimulate a chemosensillum of the silk moth.

The name of this compound is derived from the Latin name for the species, *Bombyx mori*, and it was the first insect pheromone to be identified. Its discovery represents a major landmark in the history of both biology and chemistry, but it was much more than the identification of a chemical compound. For the first time, the way in which insects (and other animals) communicate by exchanging mol-ecules, using a language made up of chemical signals, became evident.

Pheromones and odours

In fact, the idea that insects could find each other by following scent traces was first proposed by Jean-Henri Casimir Fabre (1823–1915), a French entomologist, but it was only thanks to the ingenuity and the perseverance of Adolf Butenandt that in 1959 a few milligrams of the first pheromone were purified from half a million female silk moths.[11]

What was at that time an enormous task and a most brilliant achievement, can now be easily replicated with our currently available analytical tools. The chemical identification of a new sex pheromone is now performed by analysing the gland of a single moth by mass spectrometry after separation on a gas-chromatographic column. The amount produced by the female moth (less than one millionth of a gram per insect) may look very tiny, but is about 100 times what is needed for such analysis.

As a result of our improved analytical tools, sex pheromones have been identified in thousands of species. Such research was stimulated mainly by the possibility of using these chemical messages to interfere with the chemical communication system, and thereby devising environmentally friendly strategies for controlling populations of pests in agriculture. It is fascinating to think that you can get rid of dangerous insects just by *telling* them to move away, using *chemical words*. This is certainly the basic idea, but in practice things are more complicated and now, more than 50 years after Butenandt's seminal discovery, we still need insecticides to protect our crops.

Besides practical applications of economic relevance, the study of insect pheromones is extremely rewarding for the scientist, because of the great diversity, high degree of accuracy, and complexity of the chemical messages exchanged between creatures often going unnoticed. In this chapter we can only get a glimpse at the hidden treasure of knowledge waiting to be discovered inside the ants' nest under our feet or in the complex architecture of a moth's antenna and its exquisite mechanism for detecting single molecules of pheromones. A more comprehensive presentation of pheromones and their intriguing aspects can be found in Tristram Wyatt's book *Pheromones and Animal Behaviour*.[12]

Sex pheromones were the first such messages to be studied and are still those better understood and most widely investigated. But, besides sexual messages, insects use pheromones to convey other types of information, such as to warn of danger, indicate the presence of food, or to compete with other males for females. But where the pheromonal communication has reached the highest degree of complexity is in social insects, for whom chemical signals are used to define castes, to assign tasks, and to recognize individuals of the same colony.

Pheromones are volatile or in some cases non-volatile compounds perceived through an olfactory or, more generally speaking, a chemo-sensory system. So, the obvious question is: in what way are they different from odours found in the environment? This point has been clarified by defining as pheromones those chemicals produced by the individuals of a given species and having an effect on individuals of the same species. Any other chemical, including pheromones from other species, is regarded and perceived as an odour. Therefore, the same compound produced, let us say, by a female moth is a pheromone for males of the same species and triggers a strong clear behavioural response in those individuals, but it can also be perceived by insects of other species as an odour to betray the presence of the insects which produced that message. It is easy to see why such chemical information is very precious for the parasites and predators which constantly and covertly monitor their prey.

In this chapter we shall look into this amazingly complex way in which insects communicate with each other.

A BIT OF CHEMISTRY

The varied structures of insect pheromones

As with smells in the previous chapters, let us look at the chemical structures of some insect pheromones and learn a bit more about these important messengers. From the simple examples shown in Figures 14 and 15 we can observe that there is nothing special about these molecules. On the contrary, they remind us of natural chemicals or the common products of metabolism. In fact, rather than creating new and dedicated tools for making pheromones, insects synthesize these important compounds using enzymes already active in the main metabolic pathways, and only introduce at the end of the synthetic chain a specific modification to make the molecule less common or even unique.

A clear example is given by the sex pheromones of Lepidoptera (moths and butterflies), represented by the molecule of bombykol. With only a few exceptions, they are constituted by linear chains of 12

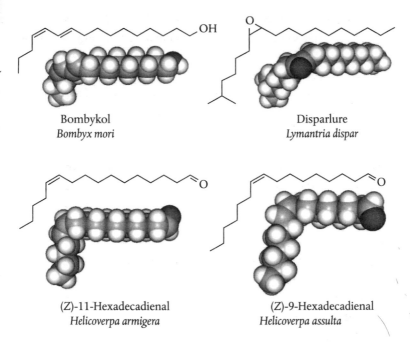

Bombykol
Bombyx mori

Disparlure
Lymantria dispar

(Z)-11-Hexadecadienal
Helicoverpa armigera

(Z)-9-Hexadecadienal
Helicoverpa assulta

Figure 14. Sex pheromones of moths. Most are linear chains, like bombykol, of 12–20 carbon atoms with a functional group at one end. One interesting exception is disparlure, existing in two mirror forms (enantiomers), one of them acts as attractant, the other as repellent for the males of the gypsy moth, *Lymantria dispar*. The two isomers of hexadecadienal both constitute the sex pheromone blend for two sibling species *Helicoverpa armigera* and *H. assulta*, but in opposite ratios, of about 95:5.

to 20 carbon atoms, often bearing one or two double bonds and a functional group at the end, an alcohol, an aldehyde, or an acetate. These molecules resemble very closely the common fatty acids found in plants and animals, only differing in the type of functional group or sometimes in the position or the configuration of a double bond. It is easy to visualize that by playing with such simple structural elements in combinations we can obtain a very large number of chemicals different enough from each other to be uniquely identified by the olfactory system of the insect.

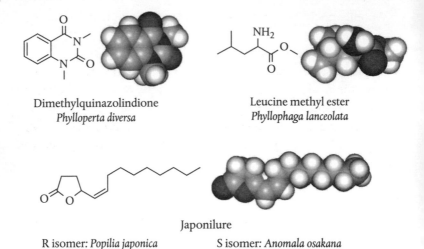

Dimethylquinazolindione
Phylloperta diversa

Leucine methyl ester
Phyllophaga lanceolata

Japonilure

R isomer: *Popilia japonica*   S isomer: *Anomala osakana*

Figure 15. Insect sex pheromones include a large variety of chemical structures. Particularly intriguing is the case of two scarab beetles, both using japonilure, a fruity smelling lactone which exists in two mirror-identical forms (enantiomers). The 'Japanese beetle' *P. japonica* uses the R form as an attractant and its enantiomer as a repellent, while for the Osaka beetle *A. osakana*, the roles of the two compounds are reversed.

We can still recognize the same type of skeleton in the molecule of disparlure, the sex pheromone of the gypsy moth, despite its strange appearance. In fact, the unusual three-member ring (an epoxide) in the middle of the molecule is easily obtained by oxidation of a double bond. In other cases, the precursor linear fatty acid can be more difficult to spot, as in the γ-lactone (Figure 15). We have already met these cyclic compounds in the flavours of roasted meat (short chain lactones) or fruit (long chain derivatives). They are no more than esters and formed when an acidic group meets an alcohol hydroxyl group. In lactones both groups belong to the same molecule and we end up with a cyclic structure. We just need to open the ring to restore the acid and alcohol groups and observe that the molecule only differs from a common fatty acid by the addition of a hydroxyl group.

Another pheromone reported in Figure 15 is clearly a derivative of an amino acid, and in particular is the methyl ester of leucine, one of the 20 amino acid building blocks of proteins. The other two examples just illustrate the variety of structures that can be found among the many chemical words that insects use to communicate within each species. The uniqueness of the messages and an accurate discrimination by the detection system are the basis for reliable recognition of a partner, thus avoiding unfruitful mating between different species.

Pheromones can be complex blends

To make the message clearer and more accurate, a sex pheromone is made up of a blend of chemicals, generally two or three major components accompanied by others in trace quantities. These complex bouquets are more like sentences than single words; or we can go back again and find a similarity with Chinese words. While in classic Chinese the rule was that a concept would be expressed with a single character, nowadays words are often the combination of two or three characters, whose individual meanings are complementary (to make the concept more narrowly tuned) or similar, thus reinforcing one another and avoiding confusion in the spoken language where it is common for the same sound to indicate different concepts.

Two noctuid moths, *Helicoverpa armigera* (the cotton bollworm) and *Helicoverpa assulta* (the oriental tobacco budworm) provide a nice example of two species using the same compounds to make their pheromone blend, but in opposite ratios (Figure 14). The two components are both linear aldehydes of 16 carbon atoms, each with a double bond, but one in position 9, the other in position 11. Each species uses a blend of both aldehydes in the approximate ratio of 95:5, with the components reversed for the two species. This capacity for recognizing minor differences in the molecular structures, such as the position of a double bond, indicates a detection system capable of fine discrimination.

Amazing similarity and subtle differences

In other cases differences can be even more subtle and the perception system is able to distinguish between molecules that are identical in every respect, except for one being the mirror image of the other. This phenomenon is called *chirality*, from the Greek word for hand, and describes molecules identical in their functional groups and in the arrangement of their atoms, except that one is the mirror image of the other, like two hands. Molecules which can exist in two such forms are called chiral compounds. We can easily verify if a molecule possesses such a property by looking at its symmetry. A symmetrical molecule is unique and, like any symmetrical object, is superimposable with its mirror image.

Asymmetric molecules are very common in biology, from amino acids to sugars and other chemicals: in most cases, only one form is found in nature. Thus, amino acids making up the proteins of our body are all of the same chirality: if we compare them, although different, all have the common part oriented in the same way: their mirror images are not synthesized. The same is true for sugars which are found everywhere from simple molecules like those in fruits (glucose, fructose, sucrose) to the long chains of starch, cellulose, and other polysaccharides.

But in some cases we find both mirror images in nature. We call these compounds *enantiomers* and they are distinguished from one another by a letter (L for left or D for right) or a sign (+) or (−) preceding the name. These notations refer to the single physical property which can distinguish two enantiomers, that is the capacity of turning the polarization plane of a beam of light to the right or to the left.[13]

When both enantiomers of a compound are found in nature, they are not usually synthesized in the same place. In other words, the synthesis is *stereospecific*, only creating one isomer in a given tissue or organism. One well known example is the minty-smelling carvone. Its L form is produced in the leaves of spearmint and contributes to the typical scent of this plant, while the other enantiomer (D) is found in the seeds of caraway, of which flavour it represents an important component. This is probably the best example of two enantiomers smelling different, although very similar, to humans. In fact, generally speaking, we are not good at discriminating such subtle differences and both enantiomers of all common odorants smell identical or very similar to us.

With insects, it is a different story. Their olfactory system can easily distinguish one pheromone from its mirror image, indicating that their olfactory receptors are very finely and narrowly tuned. This is the case, for instance, with the gypsy moth and its pheromone, which we met earlier in this chapter. When disparlure was first discovered, chemists immediately devised a synthetic method to produce this compound in bulk quantities for uses in agriculture. Disparlure can exist in two mirror image structures, but the moth produces only one form to be used as a pheromone. However, the chemists preferred to synthesize the *racemate*, that is a 1:1 mixture of the two isomers, which could be made much more easily and cheaply. They expected that the product would offer a 50 per cent efficiency, as only one enantiomer was the active compound. Instead, much to their surprise, they found that the racemate was a very poor attractant which was of little use in agriculture. The reason was that the 'wrong' isomer acts as inhibitor and suppresses to a great extent the signal produced by the active component. This mechanism very possibly allows the olfactory system to efficiently distinguish between the presence of a female of the same species and those of related species sharing the same environment.

A dramatic example is provided by two Japanese coleoptera, *Popilia japonica* and *Anomala osakana*: both species use the same pheromone, a gamma lactone of 14 carbon atoms, named japonilure (Figure 15), which smells fruity to us. This compound can exist in two mirror images and both insects use both enantiomers. Nevertheless, they can accurately distinguish the signal of their own females from those of the other species, because one of the enantiomers is an attractant, the other acts as an inhibitor, only these are reversed between the two species. All known insect pheromones are listed in a freely accessible database.[14]

There are also anti-aphrodisiac pheromones. Many examples are reported where a male, after mating, leaves a repulsive odour on the female to prevent other males approaching. Particularly curious is the case of the micro wasp, *Ooencyrtus kuvanae*, a parasitoid of the gypsy moth which lays its eggs inside the eggs of the moth. Being short-lived, the males try to use their time as efficiently as possible by mating with the maximum number of females. As the process of mating takes time, during which other opportunities can get lost, why not book the largest number of females for future sex? Adults swarm around all at the same time. Competition is very high and there is a frenetic rush to get to the females. Therefore, males have developed an interesting strategy. Rather than mating with females one after another in a rush, they mark a large number of females with a pheromone. A *tagged* female is not approached by other males and accepts only the male who left its signature on her. In this way, they quickly build a sort of harem, to which they return later to complete their job. The nature of the pheromone is still unknown, but careful observation of the behaviour has shown that it is delivered by the antennae of males onto the antennae of females. Usually antennae are sensory organs built to receive rather than emit chemical signals, but pheromone glands have been observed on the antennae of solitary bees, while the habit of exchanging information by touching each other's antennae is well documented among ants.

CURIOUS FACTS ABOUT PHEROMONES ACROSS SPECIES

Puzzling coincidences

A few years ago a scientific discovery hit the news for its apparently remarkable aspects. The sex pheromone of the male elephant was identified as a linear alcohol of 12 carbon atoms (7-dodecenyl acetate), a very common pheromone among moths, shared by more than 100 species (Figure 16). What triggered the curiosity and appeared hard to accept was that such diverse animal species could use the same molecule to send messages. In fact, there is nothing at all strange about this. The basic metabolic pathways in insects and mammals are the same and fatty acids are produced all the time in large amounts. No wonder, then, that more than one species utilizes the same enzyme to produce other chemicals, such as pheromones.

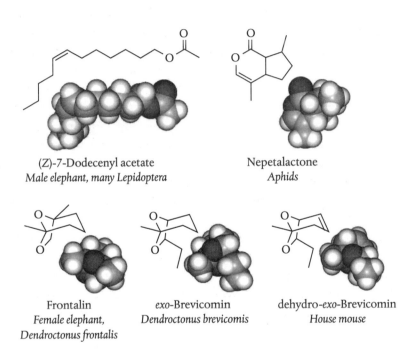

(Z)-7-Dodecenyl acetate
Male elephant, many Lepidoptera

Nepetalactone
Aphids

Frontalin
Female elephant,
Dendroctonus frontalis

exo-Brevicomin
Dendroctonus brevicomis

dehydro-*exo*-Brevicomin
House mouse

Figure 16. Curious coincidences between insect and mammalian pheromones. Dodecenyl acetate, a compound occurring in the pheromone blend of many Lepidoptera is also the male sex pheromone of the elephant, while the female elephant produces frontalin, a molecule which is also the pheromone of the Southern pine beetle *D. frontalis*. A closely related species, the Western pine beetle, *D. brevicomis*, uses as a pheromone a molecule structurally similar to frontalin, but almost identical to a mouse pheromone excreted in the urine of males. Nepetalactone is a component of most aphid species sex pheromones, but it is also, for reasons that remain unknown, a strong attractant for cats.

On the other hand, the fact that moths and elephants share the same pheromone is not going to raise any concern about the possibility of choosing the wrong partner, leading to unsuccessful mating. The real problem is, from a chemical ecology perspective, that the same dodecenol is found in the pheromonal secretions of a large number of moths, among whom the danger of errors in recognition of the partner could be very high. However, we have already discussed how insects solve this problem by using mixtures of chemicals as attractants, rather than single chemicals. In any case, if you see male moths flying around an elephant, they are certainly looking for females, although we know they are on the wrong scent track.

Equally strange is the high similarity between one of the pheromones of the house mouse and that of the bark beetle, *exo*-brevicomin and dehydro-*exo*-brevicomin, respectively (Figure 16). The difference is only a double bond. Again, as in the case of the elephant and the moth, this fact does not contain any hidden biological meaning, being merely a coincidence, which is not going to cause any problems between the two species.

Another curious coincidence puts aphids and cats in the same bracket. Before bringing in cats, aphids deserve some special attention when talking about pheromones. These tiny insects, responsible for enormous losses in agriculture, come in a large variety of species and have colonized almost every kind of plant on the planet. It is an amazing fact that most aphid species utilize the same four chemicals as components of their sex pheromonal blend, which each species mixes in a unique composition. These compounds are a delta-lactone, called nepetalactone (Figure 16) and three isomers of the corresponding alcohol, nepetalactol. These names derive from that of a common herb *Nepeta cataria*, also known as 'catnip', producing relatively large amounts of the lactone. If you grow this plant in the garden you will have witnessed the strange behaviour of cats rubbing their chin against the plant and becoming excited because of the smell. You would think that catnip acts as a drug for the cat, but we still do not know how it works.

Deceptive messages from plants

The production of the same or very similar molecules in phylogenetically distant species is certainly fortuitous, but in several cases these facts are utilized by other animal or plant species to attract insects to their advantage. Setting up sticky traps loaded with pheromones on a tree to catch dangerous species is one example of how we exploit our knowledge to deceive insects with chemicals produced in the lab. The equilibrium between the many species sharing the same habitat is often based on such deception mechanisms which have reached high degrees of sophistication across evolution.

Insects play an important role in pollinating a large number of plant species and plants have developed all sorts of strategies to attract insects—bright colours, scents, and shapes. Orchids have gone further. Many species exhibit flowers which in shape and colour resemble the solitary bees which pollinate them. But smell is more important for insects and nothing is so irresistible as the scent of the female. To make deception more efficient, some orchids, such as *Ophrys sphegodes*, release a bouquet of volatile chemicals very similar to that produced by their pollinating insects, in this case the solitary bee *Andrena nigroaenea*. The deception is so convincing that the bee is fooled into attempting to mate with the flower, in the process covering itself with pollen which will be discharged in another flower.

The case of the arum and the blowfly is similar, although we cannot speak of pheromones in this case, only of attractants. The mediterranean 'dead-horse arum' (*Helicodiceros muscivorus*) produces a very large and beautiful flower, which however releases a stench very similar to that of a rotting carcass. Blowflies find this bouquet very attractive and visit the flower, which also provides its guest with a warm environment. Probably the combination of temperature and odour is effective in fooling the flies, who wander inside the flower in search of food. As with the bees and the orchids, pollination is assured without any reward for the insects.

A more finely tuned mechanism to attract pollinators is that adopted by ancient parasitic cycad plants growing in the Australian desert. These plants occur as distinct male and female individuals and thus have to face the problem of sending their pollen from one plant to another. As with many other plants, Australian cycads attract some primitive insects of the genus Cycadothrips to collect pollen. β-Myrcene, a hydrocarbon, smelling herbaceous and balsamic to us, is a pheromone for the thrips and is produced by these plants to attract their pollinators. But Australian cycads have developed a special mechanism to make pollination particularly efficient. During the day the flower can increase its temperature by up to 12 degrees, thus driving the thrips away. However, it is not the heat that makes the insect uncomfortable, but rather the same smell. In fact, the same β-myrcene, very attractive at low levels, becomes aversive when its concentration increases due to the rise in temperature. Female flowers apparently do not exhibit this mechanism, as their temperature remains more or less constant. Consequently, the insects driven away by the strong smell of male flowers are again attracted by the female flowers, thus completing the pollinating process.

Mating and death

Following the sex scent in other cases might be life threatening for some moths, as in the popular Chinese story of the beautiful girl appearing to a traveller in the night and then turning into a fox (the fox is an evil animal in Chinese tradition). Bolas spiders are so called because, instead of weaving a web they produce a long thread of silk with a little sticky ball at the end, which they swing like bolas trying to fish moths flying nearby. The moths are attracted by chemicals reproducing their sex pheromones, with which the bolas are impregnated. Usually, each species of these spiders hunts a single species of moth and therefore only produces one type of pheromone. However, at least in one case, the spider makes a mixture of two pheromones for two species of moths, which fly during the night at different times. To optimize catches, the composition of the blend changes during the course of the night according to the change of moth populations. How evolution has adjusted the synthesis of specific chemicals and induced the establishment of such robust and sophisticated behaviour is hard to understand. The ecological system is very finely balanced and the presence of bolas spiders does not put much pressure on the moths to disrupt this chemical communication, nor is the level of preying so excessive as to risk local extinction of the moths.

PHEROMONES AS A CHEMICAL LANGUAGE

The complex communication in insect colonies

If you think sex recognition between insects is amazingly complex and sophisticated, take a look at social insects, among which phero-mone communication is much richer and chemical words are used in a sort of language to exchange various types of information.

Colonies of social insects, such as honey bees, wasps, ants, and termites contain huge numbers of individuals divided into castes, each with specific tasks to accomplish for the survival of the colony. If you have ever looked inside a bee hive or watched ants busy carrying food to the nest, you can immediately perceive that each individual is moving according to specific orders, like workers in a factory or in a work camp. To achieve such efficient organization one would expect there to be a boss supervising all activities and making sure every individual acts in accordance with its specific task. Worker honey bees perform different operations during their lives: feeding the larvae, tending the queen, cleaning the cells, guarding the hive, and of course going out to collect pollen and nectar.[15] Who is going to tell any single bee what its job should be for the day? Termites build huge and complex mounds (on a human scale it would be like our highest skyscrapers) with ventilation corridors and an intricate maze of chambers, without either a supervis-ing architect or a plan drawn to scale. The galleries made by ants underground are no less complex and organized.

The ant colony as a superorganism

Several scientists have proposed the idea of *superorganisms* in relation to organized communities and even to the whole world. E. O. Wilson and Bert Hölldobler have dedicated their lives to the study of ants and are the best authorities in this field. They have applied the concept of superorganism to insect communities and have greatly contributed to popularizing the complex life of ants.[16]

The idea that a large society of individuals, like ants or bees, could be regarded as a superorganism hypothesizes that the tight connections between individuals would produce a robust and reproducible network, so that the community acts in some way as a single organism. It would be rather like the cells in our bodies, which have been differentiated to perform specific tasks, but are still continuously talking to one another. Or like the neurons of the brain, all connected and working in synchrony. Single neurons can perform very simple tasks, but many linked in a network can solve complex problems.

The cells of our body or the neurons in the brain differentiate under specific orders coming from inside, often in response to external stimuli, which switch on the expression of some genes, while suppressing the activity of others. Neurons are activated or inhibited by chemical signals released by other neurons and the complex firing propagating through some specific paths of the entire network in the end produces a behavioural effect. In a colony of social insects information is continually exchanged using chemical signals, namely, pheromones, which have precise meaning and switch on individuals into performing some established and predicted actions.

Can we consider a colony of ants as a superorganism? The question has been debated for a long time and is still open. Certainly in nature there are different degrees of complexity and different levels of organization of individual entities into complex organisms, from the cells in our body, or within an organ, to the ants in a large nest. There are also intermediate cases, like slime moulds. These are single-celled organisms each living its own independent life as long as food is abundant. When food becomes scarce, cells can get together to form a single organism, capable of moving around and detecting food through chemoreception mechanisms. The cells arrange themselves in different shapes, such as stalks and fruiting bodies, which eventually produce spores. Slime moulds are amazing examples of how single cells can organize themselves into complex organisms capable of new functions, in response to environmental pressure.

What is certain is that all such complex behaviour, from the cells of our brain to slime moulds and insect colonies, are all the effect of chemical communication. We have so far discussed sex pheromones, which regulate behaviour between the sexes: search, courtship, and mating. But as the life of insects is more complex and involves interactions between a large number of individuals, rather than a simple relationship between a male and a female, we need more pheromones to convey information and send orders.

Chemical communication in social insects

Thus we have alarm pheromones to warn other members of the colony of dangers and other pheromones showing the way to good food sources. Pheromones also mark each individual with the caste and nest to which they belong. When two ants meet on a trail they sometimes stop and smell each other, touching their antennae together. They might exchange information about food sites, but they also recognize each other as members of the same nest or as foreigners.

Honey bees possess several glands that secrete a variety of pheromones. The 'queen pheromone' is very efficient in inhibiting maturation of the ovaries and as long as there is a queen in the hive, workers do not lay eggs.[17] But as soon as the queen dies, this inhibition is removed, so that one of the workers becomes the new queen, thus ensuring survival of the colony. For this mechanism to be efficient, the olfactory message should be volatile enough to disappear shortly after the queen dies. In fact it is constituted by a mixture of compounds of medium size, the main one being a chetoacid of 10 carbon atoms (Figure 17), together with its corresponding alcohol and two aromatic compounds. This blend is continuously collected by workers from the queen and handed over to all other workers in the hive. With such a continuous supply of the queen pheromone, all the workers are kept under control, but soon after she dies this smell has to be reduced to

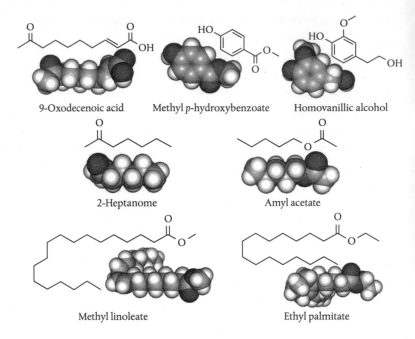

Figure 17. Honey bees use a large variety of pheromones to control the life of their large communities. In the top row some components of the queen pheromone, a mixture of chemicals released by the queen in its mandibular glands and essential to prevent workers from laying eggs, among other functions. In the middle two small molecules released in situations of danger which, owing to their volatility, can be rapidly eliminated when the emergency is over. The last two long-chain esters are components of the brood pheromones, released by the larvae to call the workers to their tending duties.

levels low enough for ovulation to be restored in the workers. These same chemicals, together with another aromatic compound, coniferyl alcohol, and additional minor components, constitutes the so-called retinue pheromone, responsible for keeping a troupe of maids around the queen.

Another scent marker attached to her eggs by the queen ensures that only the larvae emerging from such eggs are reared. In fact, eggs occasionally laid by some workers and lacking this pheromone are rapidly removed and discarded. Such eggs would give birth to males, useless in the normal life of the colony. However, when the queen dies males are needed to inseminate the new queen.

Other secreted chemicals act as alarm pheromones. These are also volatile compounds and have to be dispersed in the air as soon as the danger is over. When a bee stings an animal it releases a sweet smell of banana: it is a complex mixture, whose main components are amyl acetate (Figure 17) and other esters. This pheromone recruits other bees with the order to attack and sting. Due to its high volatility, the pheromone travels and acts quickly, but also disappears soon after the situation is back to normal. There is another alarm pheromone, secreted by a different gland, also a volatile compound: 2-heptanone (Figure 17). This has the effect of sending away intruders and is also used to keep bees away from unrewarding sites. Thus, by modulating the secretion of different compounds, bees, like other social insects, can finely adjust the meaning of the message, while the volatility of the released compounds ensures residence of the message in the environment for the necessary length of time.

Larvae also secrete their pheromone. This is the so-called brood pheromone, constituted by a number of fatty acids methyl and ethyl esters, such as ethyl palmitate, methyl linoleate (Figure 17) and other similar compounds. These chemicals have reduced volatility, so they are active for relatively long periods, prompting the workers to feed and take care of the larvae. And there is more about these pheromones. They convey information about the age and health of larvae and send instructions to the workers regarding their specific duties.

Besides this complex and highly regulated system of chemical communication with pheromones like written orders and rules to obey, we should not overlook the role of general olfaction in social insects. Honey bees are confronted with a large variety of smells and communicate to the other members of the hive what they have experienced. Returning home, these bees bring in the scents of the flowers they have visited, which fill the hive and are rapidly fixed in the memory of the other workers. The well known dance of the bees[18] to indicate places of foraging is accompanied by frequent interruptions, during which the bees who have brought back pollen and nectar offer what they have collected to those who stayed at home. Often such interactions are so brief that the potential recipient is not able to receive the food, but only to smell it. So, information about the foraging sites encrypted in the pattern of the dance is enriched by olfactory cues, as the other bees are given the opportunity to smell what they are expected to find in the new locations. It is like guiding someone to a new place by providing maps and directions, but also showing pictures of the site so that the place can be recognized with confidence. Bees use olfactory images instead of pictures. Such complex performances rely on the unique capacity of the honey bees to learn new smells and store such information in their memory with high accuracy.[19]

Ants have also been the object of much observation and research, regarding their chemical communication. As is the case with honey bees, ants are organized in hierarchical colonies with a queen, foragers, soldiers, and other workers. Again segregation between castes and assignment of tasks is regulated by pheromones. Unlike honey bees, ants do not generally fly, but walk to their food sources and are therefore able to mark long trails with their pheromones to show the way to other members of the same nest. These scent markers are rather volatile and do not persist for long periods. While there is food at the site, workers tread the path and in doing so refresh the scent by releasing more pheromone from their legs. When the food site is fully exploited and workers no longer pay visits, the trail pheromone rapidly disappears, thus avoiding unfruitful trips. Moreover, to make sure that unrewarding sites are not visited any longer, ants put an additional 'no entry' signal to paths that are not going to be used again.

It is also interesting to understand how the trail pheromone helps ants to choose the shortest path between two sites. We can imagine that at the beginning ants might try different routes. While walking to the food source, they mark their path with pheromones up to the food source; then they return to the nest along the same path which they have marked, again releasing more pheromone. In this way, the shortest path receives more pheromone. Therefore, the shorter the path, the stronger is the smell. Ants simply follow the strongest signal.

A chemical citizenship

If an individual coming from another colony tries to enter a nest, it is immediately recognized as an enemy and chased away or killed. Most wars occur between ants' nests and sometimes individuals of the defeated colony are taken prisoner and brought to the victors' nest. Then, fast and extensive rubbing occurs between the members of the colony and the new arrivals, who are thus anointed with the smell of the new colony and recognized as members. It is like a citizenship granted to the prisoners, who receive an odour marker as an identity card. This bouquet is generally a mixture of long-chain hydrocarbons, the so-called *cuticular hydrocarbons* being produced by cells just under the cuticle. This scent has to last a long time and is therefore made of heavy hydrocarbons of 20–40 carbon atoms. Most of these compounds are solid, a kind of wax, and it is not quite appropriate to think of odours, as they are not volatile enough to reach the antennae airborne. We could refer to a sort of taste perception, or more accurately, to contact chemoreception.

In fact, in insects distinction between smell and taste cannot be based on the organ of perception. Mammals smell through the nose and taste with the tongue. Insects smell mainly with antennae and taste with their mouth organs, as well as with their legs; but the picture is more complex than that. Olfactory sensilla and taste sensilla, anatomically distinct, can be found in all parts of the body, including the ovipositor and the wings.

Not solitary, yet not quite social

Aphids are not social insects, as they are not organized in societies, but they live in large groups and therefore sometimes need pheromones to talk to each other. Besides sex pheromones which we have already introduced, basically the only important message to convey is the presence of danger, and aphids produce a very potent alarm

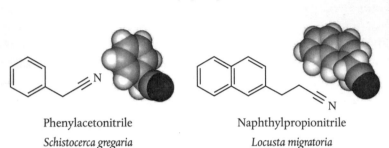

β-Farnesene

many aphid species

Phenylacetonitrile

Schistocerca gregaria

Naphthylpropionitrile

Locusta migratoria

Figure 18. Aphids and locusts are not social species, but live together in large groups. Several species of aphids use β-farnesene as the alarm pheromone, a sort of common word across different languages. The two nitriles, uncommon in nature, are produced by locusts as aggregation pheromones and courtship deterrents.

pheromone. This is released when a predator is approaching or when there is a life-threatening situation. Aphids respond immediately to this message by allowing themselves to drop from the leaves in order to return to their food source when the danger has ceased. β-Farnesene (Figure 18) is the compound used by most species of aphids for this purpose and therefore the same message is shared by individuals across many related species. Being a hydrocarbon, β-farnesene is quite volatile, an important condition for ending the alarm as soon as the danger has disappeared. The rapid termination of the message is further assured by the chemical instability of the molecule, particularly in the presence of sunlight.

Some plants, such as the wild potato, have evolved to produce β-farnesene as a product of their complex secondary metabolism. This fortuitous fact has granted these plants immunity against attack by aphids, providing an additional evolutionary advantage. Needless to say, scientists have produced transgenic plants able to synthesize β-farnesene with the aim of protecting valuable crops from the devastating effects of aphids and the diseases they carry.

Another case of insects who cannot be considered as social, but who nevertheless aggregate and live together, are locusts. We are all aware of the wide destruction they bring to crops. When a swarm of locusts arrives there is no remedy, it is a real plague as detailed in the Bible, and when they leave, what is left behind is a desert.

Locusts undergo a curious physiological change during their life. When they reach adulthood they are solitary, if they have been reared in the proper way. This means that they do not gather to form swarms. But, when put together, they start chatting to one another and become very social, without however organizing themselves into a structured society. This is a real physiological shift accompanied by a phenotypical change of colour from green to brown in the case of the oriental locust *Locust migratoria*, and a series of modifications at molecular level. Recently, DNA methylation has been shown to play an important role in such a transition. It is in the gregarious phase that they become aggressive and devastating. Therefore, preventing such a physiological shift would drastically reduce the damage of these insects to crops. The element responsible for initiating a complex series of biochemical events leading eventually to swarming has been identified as a volatile molecule, at least in another economically important species, the desert locust *Schistocerca gregaria*. It is a simple but unusual molecule, phenylacetonitrile (Figure 18) which has been classified as both a gregarization pheromone (because it induces the phase shift) and an aggregation pheromone (because it causes the aggregation of gregarious locusts). However, the same compound, at relatively high concentrations, acts as a courtship-inhibitor to avoid homosexual attacks. Similar compounds, bearing the uncommon nitrile group linked to an equally uncommon naphthalene system, have been identified in the oriental locusts (*Locusta migratoria*), where they are likely acting as pheromones.

We are slowly decoding the chemical language used by locusts to aggregate, a first step towards interfering with their communication system and eventually preventing swarming. At the same time, however, as our knowledge improves, we become aware of unsuspected complexity, where the same chemical may induce different behavioural responses depending on its concentration, the environmental context, and the physiological conditions of the insect.

LEARNING FOREIGN LANGUAGES

Mastering a foreign language certainly gives us advantages. If we are able to read the signs we can find our way to the supermarkets and avoid starving while in a foreign country. If we understand warnings of danger approaching, we can find a safe place and save our lives. During war, it is fundamental that spies understand the language of the enemy.

Smelling the prey

Some insects have also developed a skill for understanding and sometimes even speaking the languages of other species. A well known case is that of ladybirds, who, despite their attractive appearance, are ferocious predators. Their preferred delicacy is aphids and they have learned to smell their prey through the alarm pheromone, which is very similar or even identical across many species of aphids. β-Farnesene is a molecule we met earlier, which is secreted by aphids when in danger, to warn their comrades. It is curious that the very compound produced to avoid danger also has the effect of attracting the predator and making the situation even worse.

The β-farnesene produced by aphids can also be detected by ants to locate aphid colonies, with which they establish a relationship of mutual advantage. In fact, ants are very keen on the sugar secretion aphids produce and they make certain of keeping a reliable herd of aphids to farm. How do ants accomplish this task? It has been observed that chemicals released by ants from their feet and known as trail pheromones are also used to tranquillize and subdue aphids. In this case, ants can both enslave aphids and also provide some advantage by keeping them away from ladybirds, thus protecting their honeydew supply.

The habit of eavesdropping is quite common with parasites and parasitoids. Several species of wasps inject their eggs into the larvae or into the eggs of moths. A tiny wasp, *Trichogramma brassicae*, which parasitizes the cabbage butterfly *Pieris brassicae*, can smell the odour of mated females and jump on them, ready to colonize their eggs as soon as they are laid. The smell detected by this wasp is an anti-aphrodisiac odour, phenyl acetonitrile (by coincidence, the same aggregation pheromone for locusts we have just met above), which the male puts on the female after mating to make her less attractive to other males.

The brood pheromone released by honey bee larvae, as a request to be taken care of by the workers, also turns out to be a weapon against the same larvae. Their cries for help are detected by parasite female mites of the genus *Varroa*, which enter the cells about one day before they are capped and so are sealed in with the larvae. The mites then lay eggs and feed on the larvae. By the time the adult bee emerges from the cell, several of the mites are completely developed, have mated, and immediately start searching for other larvae to parasitize.

OF MEN AND MOSQUITOES

Wherever you go in the world you find mosquitoes. From the tropics to the arctic, they have adapted to all climates, so long as there is water and animals from which to suck blood. We all know how annoying they are and what a nuisance the itching they cause can be. But this is nothing compared to the serious diseases they carry worldwide, particularly in tropical countries. Malaria, dengue fever, yellow fever, and chikungunya are some of the worst and are responsible for millions of deaths each year.[20]

Mosquitoes need blood

To lay eggs mosquitoes need a blood meal to complement their otherwise vegetarian diet. How do they find their hosts? Smell, of course, but their strategy is a bit more complex. They actually follow three cues to find a live animal: temperature, carbon dioxide, and smell. The first two elements tell the insect that their host is alive and breathing, while different smells can provide more information about the host. In fact, mosquitoes are quite choosy and different species or even different populations prefer pigs, mice, birds, or humans.

Repellents to replace insecticides

Research aimed at reducing the population of mosquitoes has been boosted in recent years. There is also a compelling need to replace insecticides with alternative strategies, such as insect repellent for use on the body or in closed areas. For a few decades repellents based on DEET and Icaridin, two synthetic cyclic amides (Figure 19), have been employed with some success, although only when used in high doses. In fact, different formulations of spray lotions, creams, or sticks contain not less than 10–20 per cent of the active compound in some cases even 40 per cent. Such disturbingly high doses and the recent concern about neurotoxic effects of these compounds, which bear structural similarities to some insecticides, has prompted the search for alternatives.

Natural compounds, all components of plant essential oils, have received much attention based on the misleading idea that what is natural is safe. Citronella scented candles are commonly used, but also body sprays and lotions for the body containing herb extracts are commercially available. Hundreds of plant extracts, as well as their components, have been found to be active against mosquitoes, but again the concentrations needed are quite high, never less than a few grams for every 100 mL.[21]

Menthol, thymol, and eugenol, all natural compounds we met in Part 1 of this book, as well as oils of sage, thyme, oreganon, and many

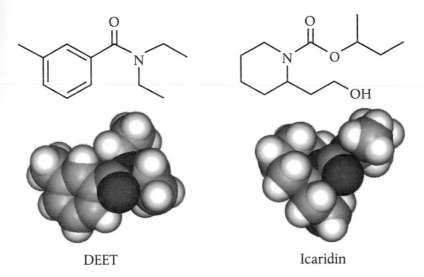

DEET   Icaridin

Figure 19. DEET and Icaridin are the most commonly used mosquito repellents. Concern about their neurotoxicity has stimulated much research towards naturally occurring repellents. In fact, most essential oils and their constituents, such as citronellal, menthol, thymol, cinnamates, and many other compounds present in flowers, herbs, and cooking spices are as effective in keeping away mosquitoes and other blood-sucking arthropods.

others are all good repellents, when used in high concentrations. It is again a curious coincidence that among the most active natural repellents we find the oil of catnip, which is commercially available for such uses, thanks to relatively high levels of nepetalactone, the aphids' sex pheromone we discussed earlier in this chapter. Unfortunately, the repellent activity of these oils rapidly decreases with concentration and many natural compounds, reported as repellents at high levels, become attractive when their concentrations drops below certain limits. That the smells of herbs and flowers, such as citral, linalool, geraniol, and menthol could be attractive to mosquitoes is not surprising. In fact mosquitoes feed generally on nectar; only females, and then only before laying eggs, have to supplement their diet with the proteins they get from a blood meal.

It is rather disappointing and frustrating to think that a repellent we use on our body can turn into an attractant after a certain time. However, the dual and contrary effect of the same odorous compound is not surprising. We have already considered such situations when describing some food flavours—pleasant at the trace amounts present in foods, yet repulsive when their concentration is raised above a certain level. The clearest example was that of linear aldehydes, as nonenal and nonadienal, which contribute a fresh quality to cucumber, but become repulsive at high concentrations when produced during degradation of fats. We must simply accept that insects are not simple machines to be switched on and off at will. The processing of olfactory signals in their admittedly tiny brains is quite complex and sophisticated and the connection between peripheral sensory hairs and the brain is by no means direct and well understood.

Messages of danger or confusion?

At this point, in order to proceed further in our search for mosquito repellents we now have to ask a basic question: what is an insect repellent? There are two possible answers to this. Broadly speaking we can call a repellent anything that is efficient in keeping away mosquitoes. But if we want to attach a biological meaning to this word, we could define a repellent as any chemical carrying a message of danger. The best example fulfilling such a restricted criterion is β-farnesene, the alarm pheromone of aphids. This compound, synthesized by the aphids, is released with the specific purpose of warning other aphids of a dangerous situation. Therefore, it can be defined as a pheromone and, like all pheromones (vital messages) it is perceived at extremely low concentrations. Of course, the definition of pheromone for β-farnesene only applies when aphids are the recipients of the message. For other species it is an odour, conveying specific and sometimes useful information, as is the case of ladybirds, which feed on aphids and find their prey on the trail of this scent.

From this perspective, is there a repellent for mosquitoes? Certainly we are not aware of any chemical produced by mosquitoes in situations of danger, a sort of alarm pheromone. In fact, we suspect that such an alarm pheromone does not exist for mosquitoes. So far, we have only found alarm pheromones in social insects or in species in which a large number of individuals live together, aphids for example, although without organized society. In such cases, an alarm pheromone gives out signals as it warns other individuals of the same species of approaching danger. In mosquitoes, or in other insects where individuals do not interact, except for mating, there would be no use for an alarm pheromone.

In some situations, however, danger signals could perhaps come directly from predators and enemies. While we cannot exclude such a possibility, there are no reports of such compounds, nor of any specific enemy against which mosquitoes might have developed definite defences or ways of escape.

If we want to try to make sense of the repellent effect of a very large number of natural compounds, it is first worth considering that, given a wide diversity of chemical structures, it is unlikely that specific danger signals are involved in the case of mosquitoes. Rather, we would think that such smells, used, don't forget, at concentrations thousands or even millions of times higher that those present in a natural environment, may cause confusion and mask the natural odour of the host. In any case, the mosquito senses an unfamiliar situation and probably keeps away because it is not interested, rather than being scared off.

The search for longer lasting repellents

While the search for new and more efficient repellents is going on, particularly among natural compounds, another approach looks promising for improving our currently available products. The leading idea is to modify the molecule of a good repellent, such as citronellol or menthol and make it less volatile. While this is not expected to improve its activity, less volatile products will certainly last longer. Besides, such products, not being identical with their natural precursors, are less likely to be perceived by mosquitoes as attractants when their concentration drops.

Finally, it is worth looking at the other two sensory elements used by mosquitoes to locate their host: temperature and carbon dioxide. It is likely that an integrated approach to disrupt the mosquito's sensory system might be more efficient in controlling these dangerous insects, rather than a strategy only taking into account the host's smell.

CONCLUSION

When insect pheromones were first discovered, they were regarded as magical tools that would allow us to switch insects' behavioural reactions on and off and manipulate them like tiny robots. We have learned since then that insects are highly complex organisms and perceive the chemical environment with tools much more sensitive and specific than any analytical instrument. Fooling them with our chemicals is not an easy task, as demonstrated by the poor perform- ance of pheromones to keep away pests from crops and the failure of mosquito repellents. From a chemist's point of view, insects are exceptionally good at synthesizing challenging molecules (such as chiral pheromones and complex hydrocarbons, which would take a skilled organic chemist months of work). They are astute as well at analysing complex mixtures and identifying the critical component among thousands of other chemicals, as is often the case in environ- mental situations.

Social insects with their rich and complex repertoire of phero- mones released and cleared with perfect timing regulate the life of the hive or the colony like clockwork and cope with different, even unexpected, events. The organized work of thousands of bees or millions of ants, regulated by a network of chemical messages con- tinuously exchanged between members of the colony, or the con- certed actions of termites building huge nests equipped with chambers and ventilation corridors strongly suggest that insects of a colony behave as cells of an organism and their single actions are combined to achieve tasks of higher complexity. With such a perspective the individual members of an insect colony are far from being like the gears and screws of a clockwork mechanism. There is plasticity in the performance of such superorganisms, with different tasks being assigned during the life of the colony to cope with needs originating from the development of events. Understanding how all this is achieved could probably shed light on some basic but still poorly understood strategies regulating the function of our own brains, with each neuron performing its role, but at the same time capable of adaptation and plasticity as a whole. And obviously artificial intelligence would be the next to benefit from such knowledge on the way to design new types of computers capable of much flexibility, adaptation, and plasticity, like the brain of a community of social insects.

MAMMALIAN PHEROMONES

Smelling Ranks and Kinship

Pheromones have been described in all classes and orders of animals across the phylogenetic tree, from worms to fish, reptiles, birds, and mammals. Even in primitive organisms such as bacteria and yeasts, mating is mediated by secreted chemicals, which we can reasonably classify as pheromones. Birds have been regarded for a long time as animals relying on vision and sound to find mates rather than pheromonal communication. This is certainly true to some extent, justifying the beautiful coloured plumage of many species, their courtship dances exhibiting visual displays, as well as the variety of melody in birdsong. However, evidence has recently been accumulating to show that at least some species, notably some aquatic birds, may produce and make use of pheromones for communication between the sexes. On the other hand, olfaction plays a major role in other ways in the lives of birds, for example the orientation of homing pigeons, shearwaters, and other seabirds. In this chapter we will focus on mammalian pheromones and examine a few examples, which have been studied in detail.

While there is no question about the existence of pheromones in insects and their robust behavioural responses, the field of mammalian pheromones is still an area where we should proceed with caution when analysing and interpreting reactions to chemicals in terms of pheromonal communication.[22] Although there is almost general consensus that certain stereotyped behaviours in some species are triggered by volatile compounds produced by individuals of the same species (and therefore falling into the definition of pheromones), a few scientists still disagree with such a view.[23] Their main argument is that the brains of mammals are much more complex than those of insects and that the response to any type of signal is not direct and mechanical, but is more likely to be the result of complex processing, also involving other sensory inputs. While this is certainly true and the link between pheromones and behavioural response becomes thinner and weaker in complex animals, in several cases we observe clear and emphatic responses to chemicals produced by mammals, which might confidently be classified as pheromones.

First, pheromones, unlike all other odours, are species-specific. This means that a certain molecule or a certain mixture is produced by all individuals of a species (with limitations regarding sex, age, rank, and physiological condition, but not individuality) and triggers the same response in members of the same species.

Second, mammals, with few exceptions, possess a well developed vomeronasal organ, a second nose dedicated to the perception of pheromones. The nerve endings of this organ project to, and are processed in, an area of the brain (the accessory olfactory organ) distinct from the area in which olfactory stimuli are processed (the main olfactory organ). But the distinction is not clear cut, as some pheromones are detected throughout the principal olfactory system.

We certainly can't apply to mammals the same criteria used to identify insect pheromones and expect a mechanical response, identical in all individuals and in any environmental situation. But, insofar as a specific chemical signal (produced within the species) triggers the same kind of behaviour in individuals of the same species or a subset of individuals selected by sex, age, or physiological condition, we can confidently classify this chemical as a pheromone, even though we do not expect that the behavioural response would be identical in all individuals, but might be affected by the individual situation and culture.

These considerations about whether we should classify a particular stimulus as a pheromone or a smell may seem idle and purely linguistic subtleties. But different nervous pathways are followed by the signals during processing in the two cases, which lead eventually to more or less mediated responses. The question to what extent we can stretch the definition of pheromones and whether we can apply the same criteria across the evolution of complex animals will come to a head when we come to discuss the existence of pheromones in humans.

For the time being, let's take a look at some examples of chemicals in the mammalian world which have been widely regarded as pheromones, and the behavioural effects we can observe.

SEX PHEROMONES

Talking about pheromones we immediately think of aphrodisiacs and attractants for a potential partner. In fact, sex pheromones are the best studied as they trigger the most specific and robust behavioural responses. We have seen how important it is for a male insect to identify and be attracted by the female of its own species and this task can be accomplished in a complex environment full of different odours mixed with pheromones from several related species thanks to a very sophisticated detection system, which is both sensitive and extremely specific.

Such strict challenges are not often required from mammals. Visual cues also play their role, together with auditory signals, making the task of olfaction less demanding. Even without pheromones, there is little chance that a mouse might try to mate with a cow or a horse. However, it is important for the female mouse to recognize the male carrying the best genes, as well as to avoid inbreeding. Sex pheromones in mammals are still important, although used in ways somewhat different from insects.

Mice and rats smell nutty

Pheromones of mice and rats are probably the best studied, as both species have long been models for research in biochemistry and genetics. In these rodents, urine is a major vehicle of chemical information. Whoever has been close to a mouse cage, whether in a lab or if you keep mice as pets, can immediately recognize the typical sort of nutty smell, strong, but not really unpleasant when its intensity is not too high. There are a large number of volatile chemicals in the urine, carrying a wealth of information not only on the sex of the individual, but also on its social rank and family tree. The experienced nose of the female mouse can establish, from the olfactory markers of this aphrodisiac bouquet, how long it has been since the urine message was left and whether it is likely that the intrepid male might still be around.

A volatile chemical present in the urine of mice, which accounts for its green, nutty smell is 2-sec-butylthiazoline, very similar in structure to some thiazoles we met in Chapters 2, 3, and 4, responsible for certain characteristic aromas of foods. This is a potent attractant for the female, but there is more to this volatile chemical. Its molecule is *asymmetric*. This means that we can draw two structures identical in all their parts, but for the fact that the one is the mirror image of the other. We have already met this phenomenon, chirality, when talking about insect pheromones and observed how fine the discrimination in the insect's olfactory system can be. In the mouse only the S-form of this thiazoline is synthesized and excreted with the urine. Interestingly, the synthetic mirror image of this compound (the R-form) produces in the mouse a completely different behaviour from that elicited by the natural compound.

The smell of urine marks also changes with time, as some components are more volatile than others and evaporate faster. These changes are important signals for the female, which needs to know whether the male who left this mating message is still around and available. But it is not only that. The female is also interested in the urine itself, in particular in some proteins present in exceptionally high concentration in the urine. These proteins, called MUPs (major urinary proteins) are similar or identical to those found in the nose which trap and detect pheromones. We will look at the structure and unique characteristics of MUPs later, but for now the point to note is that they are at the same time carriers of the volatile pheromones and also pheromones themselves, and can accelerate the onset of puberty in young females when they come into contact with the vomeronasal organ.

What the female mouse is looking for is this protein, which, not being volatile, cannot be perceived through the nose, but whose presence is signalled by its volatile ligand. Following this scent, and after making sure that the scent is fresh, the female finds the urine and licks it, sending the important proteins through the palate and into the vomeronasal organ.

But of course volatile pheromones produce effects on their own. Another important sex pheromone present in the urine of male mice is dehydro-*exo*-brevicomin. Together with the thiazoline just discussed it makes the urine of males very attractive to females. Both compounds are detected by the vomeronasal organ at very low concentrations, just a few parts per trillion, and in a highly specific way. Incidentally, you may remember that dehydro-*exo*-brevicomin is different by only a double bond from the bark beetle pheromone described in Chapter 5 (Figure 16).

Looking at the mechanical, almost robotical response of insects to pheromones we may get the wrong impression that these molecules might act as switches turning the insect on and off and guiding it like a toy aeroplane. Even with insects we have seen that this is not the case, as sometimes the same chemical message triggers different responses depending on other sensory inputs, the physiological condition, or the environment. With mammals we can easily imagine that the situation is much more complex. Attraction by the other sex is not only the effect of a single chemical or a simple mixture, and not all individuals of the opposite sex are perceived as equally attractive.

Repulsive to humans, but an aphrodisiac for pigs

The boar's pheromone, androstenone, is probably the best known among mammalian sex attractants and we have come across this molecule more than once in different contexts. This compound, whose strong smell is perceived as disgusting to humans (at least by that half of the population able to detect it), is a potent aphrodisiac for the sow and has the effect of making her more relaxed and receptive to courting males. The product is commercialized in spray bottles, used by pig farmers to check the best times for artificial insemination of sows.

While a pheromonal effect has been clearly demonstrated in pigs, the same molecule can also elicit strange effects in other species, as John McGlone, a professor at Texas Tech University serendipitously discovered. To stop his dog from barking he tried spraying some androstenone on its nose from a can he was using for his research. The calming effect was immediate and now the product is widely sold for this secondary use. Nobody knows how it works, but it does. As androstenone is derived from the hormone dihydrotestosterone, it is a sort of label for males across species and might produce similar or other unsuspected reactions in different animals.

The captivating scent of musk

The musk rat and the musk deer possess specialized glands, which produce pleasant-smelling compounds acting as sex pheromones for these species. These chemicals, which we have already met in Chapter 3 (Figure 10), are unusually large rings of 15–16 carbon atoms, cyclopentadecanone and muscone, respectively. Although the site of production of these molecules and some observations firmly suggest that they act as sex pheromones, detailed behavioural studies are lacking. The alluring smell for humans, immediately harnessed by the perfumery industry, was the subject of a long period of active research in synthesizing all sorts of chemicals endowed with similar olfactory characteristics. The added value of these fragrances was the belief, based on pure fantasy, that being aphrodisiacs for the rat and the deer, the same compounds might be endowed with magical attracting properties for humans as well.

Elephants ... and insects

Two sex pheromones have been described in elephants, both mentioned in Chapter 5 (Figure 16). Frontalin is a molecule of rather complex architecture and is secreted by the temporal glands of males during *musth*, a physiological state related to reproduction. The other pheromone is a much simpler molecule, the linear ester dodecenyl acetate, released in the urine by females in *oestrus*. Curiously, both compounds are also used as pheromones by some insects. The latter acetate, as we have already observed, is a component of sex pheromones of a very large number of moths, while frontalin is also the pheromone of the Southern pine beetle (*Dendroctonus frontalis*), one of the most destructive insects of pine forests in the United States.

Frontalin is an asymmetric molecule and is synthesized in different ratios between the two mirror images, depending on the elephant's age and stage of musth. Thus, by adjusting the proportions of the two enantiomers, the message can be finely tuned and adapted to different situations. We have observed a similar phenomenon with the thiazoline of mice, also an asymmetric molecule, and several cases with insect pheromones. It seems that such specificity and accurate tuning, as obtained with asymmetric molecules, is characteristic of pheromonal communication, while the perception of general odours does not usually discriminate between enantiomeric pairs of smell molecules.

NOT SEX ALONE

As in insects, chemical communication is used in mammals to convey different types of information, primarily between sexes, but also to mark territory, to show aggressive behaviour, and to mediate bonding between the mother and her young.

Scent wars among mice

In the struggle for survival and to best cope with challenges of the environment, the strongest and fittest individuals should acquire an advantage in reproduction in order to avoid the propagation of defective genes. This is the basis for the fierce competition between males to win the female to which they entrust their genetic pool. Raging battles are common between males of many species during the mating season and always end with the victory of the strongest. But there are less violent wars using smells as weapons, equally effective in selecting the strongest individuals.

Jane Hurst and Rob Beynon at the University of Liverpool, UK, who have been studying mouse behaviour and chemical communication for many years, call such competition *scent wars*.[24] As birds use their songs to advertise their presence and supremacy, mice use smells to obtain similar effects and release most of their aromatic signatures leaving around small samples of urine. We have seen how these olfactory markers can attract the female and how complex such love messages can be.

But urine signatures are also directed to other males and in such cases the message can be quite aggressive. Here we generally speak of pheromones for territory marking. In fact, male mice leave many smell markers over a certain area, which becomes a sort of private territory. An intruder, particularly if it is of lower rank (according to its olfactory signature), is expelled from the territory. If the intruder starts marking another mouse territory with its own urine, the occupier quickly countermarks by leaving more urine spots close to the areas signalled by the intruder. The message is that the mouse who is more able to refresh its marks with the stronger and fresher signature is the victor. Viewed from this perspective, Hurst and Beynon regard the marking behaviour not as a way to establish ownership of a territory, but rather as a sign of strength and supremacy. Obviously the signal is an advertisement to the females. It is like an individual buying a large car and driving around in it not because he needs more space, but as a way to impress potential partners. Or else that same individual driving at top speed and overtaking other cars not because he has urgent business, but in order to establish superiority.

The interesting aspect of these stories, from the point of view of chemical communication, is that urine traces contain a lot of data, enabling a mouse to obtain information not only about species and sex, but also individual characteristics, such as social rank, age, health, and family. In addition, and this is a very important element in scent wars, urine markers contain information on the freshness of the marker itself. In fact, supremacy of one mouse over another is largely established by how efficient and prompt an individual is in countermarking.

This complex information package contains several volatile compounds as well as proteins. Volatiles for instance include among others two isomers of farnesene which are not sex pheromones, like thiazoline and brevicomin, as they do not occur in the urine of all males, but rather provide clues to the social rank of the individuals. But proteins alone can encode the very large variety of individual features, and the urine of mice is extremely rich in proteins. We will consider these interesting components in more detail in Chapter 8, after having introduced proteins in general and their role in chemical communication. However, it is worth commenting briefly here on the function of urinary proteins in individual mouse recognition. Recall that some proteins, in particular the MUPs, are not just carriers, but pheromones themselves. The same proteins contribute to the complex signature of the urine mark at least from two perspectives. As ligand-binding molecules, they increase the shelf life of the volatile compounds, which in their absence would disappear in the space of minutes. As physiologically active molecules, they stimulate the vomeronasal system, providing specific and detailed information on the individual to whom they belong.

All this complex mechanism is confirmed by the behaviour of mice, both when a female is attracted by the presence of a male and when a male is alerted by the intrusion of another male. In both cases, the volatile molecules provide the signal, guiding the mouse to the urine marker. There follows a phase of complex inspection and physical contact with the urine, during which the non-volatile components, the proteins, are taken inside the vomeronasal organ and analysed.

Mice have been studied in detail both in their behaviour and at molecular level, but it is likely that several aspects of chemical communication in other mammals follow similar complex pathways, using both volatile scents and binding proteins.

Smell and relax

An area which has received a lot of interest recently is that of the so-called *appeasing pheromones*. These are chemical signals left by individual mammals to mark safe places and situations and which are directed to other individuals of the same species as well as to themselves. The effect of such pheromones is to release stress and induce a relaxed and confident feeling, at least so it would seem, given our limited understanding of how other animals feel.

If you have a pet, you will have noticed how it shows signs of stress when brought to a new environment. Cats, which establish affective links with the house more than with the owner, become anxious and tense if they move into a new house. You may well have noticed that in such situations they start stroking pieces of furniture with their chin. They do the same with new people, touching your legs with their face. It is not a sign of affection and they are not trying to kiss you. On their chin, cats, as other mammals, have glands secreting pheromones, which they spread around during such behaviour.

Pet owners who care about the mood of their cats can buy spray bottles to mark furniture and other elements of the house with soothing pheromones. The content of such bottles is supposed to reproduce the composition of the cat chin secretion. According to customers' feedback and to a very few scientific reports, they seem to work, but the chemical nature of such pheromones is still undisclosed.

For a different reason, scientists working in animal behaviour became interested in calming pheromones in pigs. Although in some cases pigs are kept as pets (and there is no reason why these intelligent creatures should not be treated in the same way as cats and dogs), the motivation to release stress in pigs is based on economics. Anyone who has visited an abattoir and watched pigs slowly moving in a line to be slaughtered cannot be unaware of the stress produced in these animals as they watch those in front of them shrieking and trying to escape until they fall down. But do not imagine that pig farmers or butchers desire to give their pigs a more serene death. The reason behind any concern is that when the animal is stressed it adversely affects the quality of the meat. In this case there are commercial products available which aim to produce calm. But detailed chemical information is not available.

The study of calming pheromones in mammals is certainly interesting and does not have an equivalent in insects. Their effect is quite unlike those of sex or alarm pheromones, which act more like precise and well aimed commands. These chemicals, on the other hand, influence the mood and the emotional aspects of animals and can easily be confused with other general odours coming from foods or from the environment which also produce strong effects at the emotional level. However, we can always talk about pheromones, when the chemical stimuli are produced by the individuals of a species and have a generalized effect on individuals of the same species. This distinction is important when discussing the possible pheromonal nature of smells which affect our mood and emotions as humans.

An addictive drive to milk

Links between mother and offspring are among the strongest and here olfaction plays an important role in forming relationships. Newborn mammals in particular need to establish a vital connection with their mother and her milk, at a time when in many species vision and hearing are not yet functioning. So it has to be olfactory cues which urgently attract newborns to the nipples of their mother. Survival is linked to this basic instinct which has to be satisfied within a short period from birth. We take such behaviour for granted. However, what is the attractant that drives the young in such a direct and potent way? A chemical is most likely, but of the nature of such pheromones (although we can appropriately use this term) we still know very little.

Working with rabbits, Benoist Schaal at the European Centre for Taste in Dijon in France, made an important discovery.[25] He first observed that milk was a potent attractant for newborn pups and elicited a stereotyped behaviour in their sucking from the nipple. Then, he managed to isolate a small molecule of only five carbon atoms, the aldehyde 2-methyl-2-butenal, among the large variety of volatile compounds in rabbit's milk, which produced very robust, directed behaviour in the newborn. This chemical can attract new born rabbits even when deposited on the tip of a glass rod and can fool the young into desperately sucking from the rod. The effect of the molecule is as simple and direct as that of bombykol on male silk moths and acts in the same way on all young rabbits. Moreover, this seems to be an innate response not requiring any previous learning. Therefore, this scent can certainly be regarded as a lactating pheromone for the rabbit.

The importance of such a discovery can also be found in the simplicity of this stimulus, a single chemical (quite an exceptional case even among insect pheromones) and in the emphatic response, unaffected by individual or environmental olfactory cues. In mice, for instance, a similar behaviour has been studied, but the chemical message is more complex and involves components that are learned by the pups and therefore allow newborns not only to find the nipples, but also to recognize the smell of their mother. In other species of mammals, although similar behaviour has been observed, we still have no information about the chemical bouquet, probably rather complex, responsible for guiding the young on their first step in the long struggle for life.

PHEROMONES IN PRIMATES

The attraction to the nipples being so robust and powerful, and also so important for survival, causes us to ask if it is possible that in humans a pheromonal communication of this type might have been conserved? This was one of the first questions that Benoist Schaal and his group tried to answer. Working with humans and in particular with babies is not as straightforward as using mice or rabbits for obvious ethical considerations, and results are much slower to come in. However they found evidence for chemicals produced by the mother that elicited a clear behaviour of attraction from newborns.[26] A lot of different issues must be clarified before jumping to conclusions and the authors are still very cautious in describing the behavioural effect of these chemical messengers, whose structures still await identification. But, if pheromonal communication does exist in some way among humans, here we are probably as close to it as we can be.

The issue of human pheromones is complicated and should be approached from several standpoints, so Chapter 10 is dedicated to this controversial topic. Here, we can bridge the gap between rabbits and humans and search for available information in primates, from lemurs to apes.

Lemurs, are our most distant relatives among primates and, together with lorises, bushbabies, and tarsiers are classified as prosimians, to be distinguished from simians, which are divided between New World monkeys (*platyrrhines*) and Old World monkeys (*catarrhines*), these last including apes such as gorillas and chimpanzees. Another classification adopts less familiar terms such as *strepsirrhines* (which simply means wet-nosed) and *haplorhines* (dry-nosed) to distinguish between lemurs and other prosimians on the one side and all the other primates, including tarsiers, on the other.

In any case, the species which have differentiated earlier (strepsirrhines and to some extent platyrrhines) present anatomical features and behavioural elements in common with other mammals concerning pheromonal communication, while these features are absent or unclear in catarrhines. For example, a well developed and functional vomeronasal organ has been observed in lemurs and other prosimians, as well as in many New World monkeys, while it cannot be found, with few exceptions, in Old World monkeys and apes. Another important element supporting pheromonal communication is the presence of several types of secretory glands in strepsirrhines and platyrrhines, whose products are used in territory marking. An interesting aspect of the behaviour, typical of primates and linked to the presence of hands, is the deposition of scent markers not only on stones, plants and other territorial elements, but also directly, using their fingers, on the partner's body.

Lemurs, which have received particular attention by scientists, possess brachial and antebrachial glands producing different types of secretions, which are used by the animal to mark small objects. Chemical analysis of these products has revealed a large variety of volatiles, without any one of them being responsible for a clear behavioural response. Instead, it has been suggested that this complex bouquet could provide a unique signature for each individual on the basis of different ratios between the same components.

Such behaviour is absent in catarrhines, where the presence of secretory glands (mainly in the skin and specifically in some areas such as the armpit and the reproductive organs) cannot be clearly associated with pheromone release. Some exceptions, such as the mandrill, which possesses a vomeronasal organ and exhibits a marking behaviour, indicate that both the anatomy and the behaviour have undergone gradual changes as primates slowly abandoned an intraspecific communication based on pheromones in favour of other types of messages.

Improved vision capabilities brought on by the adoption of the trichromatic system and a better perception of the three-dimensional world won out over the notion of an understanding of the environment based on chemical signals. Those who still support pheromonal communication between humans and other primates have a different view on the absence of scent marking behaviour in these species, suggesting that such practice has been replaced by the habit of exchanging odours through direct contact.

In sum, during primate evolution we have witnessed a shift from intraspecific communication that is still highly dependent on body chemicals to more complex relationships between individuals of the same species using mainly visual and auditory elements, together with chemical signals, which have, however, lost the character of compulsory orders to convey information which may affect behavioural responses, but only after having been filtered and examined through reason.

In this and the previous chapters we have looked at a particular aspect of olfaction, in which smells acquire the strength and vehemence of compelling commands which individuals of the same species can then do nothing but follow. In many insect examples this is certainly the case. From one point of view pheromones seem to rob individuals of their identity and reduce them to little mechanical robots. But pheromones regulate the life of social insects in such a well organized manner that the community as a whole acquires features and capabilities beyond the reach of the individual, thus posing the hypothesis of the *superorganism*. This idea is not so weird and improbable if we apply a similar perspective only to the cells of an organism.

On the other hand, when we move away from insects to mammals we are forced to modify our idea of pheromones and accept that the action of some olfactory messages, however powerful and specific, could be affected by the contemporary presence of other chemical stimuli, as well as visual and acoustic influences. Therefore, the same pheromone can elicit responses of variable intensities in different individuals, different situations and even depending on the past history of the receiver. Memory and associations play an important role in the life of mammals and past experience can reinforce the effect of a pheromone and at the same time can be essential in establishing a specific behavioural response. The attraction of newborn mice to the nipples is learned after smelling their mother, unlike the effect in the rabbit where the same molecule causes the same innate behaviour in all newborns. Can the two situations be assimilated and both classified as pheromonal communication? There is no complete agreement on this point, and certainly the debate cannot be put simply in semantic terms, but requires a more detailed analysis on where such chemical signals are produced and how they are processed in the brain of the receiver.

The question is even more complex and difficult to analyse when, moving along the evolutionary process, we arrive at primates and humans. The distinction between pheromones and smells affecting mood, emotions, and consequently decisions becomes intricate and complex, due to the contamination of a potential pheromonal message by environmental odours as well as by visual and auditory inputs more and more powerful as we move higher in evolution. With such complex interactions, one can understand why some prefer to simplify the problem by denying the existence of pheromones in mammals and limiting this concept to chemical communication between insects of the same species.

PART 3

PROTEINS AND SMELLS

THE BIOCHEMISTRY
OF OLFACTION

Odorants Meet the Proteins

STEPS IN ODOUR PERCEPTION

S o far we have looked into our everyday olfactory experiences and
tried to put our perceptions and emotions within a rational
scientific frame. I described the many attempts to classify odours
and to relate their diverse characteristics to structural elements of
the molecules from which they are generated.

To develop these concepts and organize our experiences we compared the stimuli, that is, the molecules conveying olfactory messages,
and our perceptions, which were described using familiar terms. So
for instance we noted that benzaldehyde smells like bitter almonds,
while citronellal recalls the scent of lemon, and 1-octen-3-ol is reminiscent of mushrooms.

This approach is certainly important and useful in creating order
among the millions of our smell sensations, but we are still very
ignorant about how the information encrypted in the structure
of a molecule is translated into a unique message perceived by the
brain.

This is where physiology and biochemistry come to our aid. Specific proteins are responsible for recognizing the various complex structures of odorants and correctly translating such features into electric signals. How do proteins accomplish this task without mistakes? Biochemistry is the discipline to answer such questions and unveil the olfactory code at the molecular level. Once electric signals are generated in response to odorants, they can be amplified, processed, mixed, and sent to the brain where eventually an olfactory image is generated. Understanding how neurons are connected along this path and how they can talk to each other, exchanging information, is the task of physiology.

Now we can begin to follow the olfactory message from the volatile molecule to the elicited behaviour across the different levels of processing and the anatomical structures involved.

In humans, the olfactory epithelium is located in the upper region of the nasal cavities. It is constituted by a layer of olfactory neurons, all aligned along the thickness of the epithelium and interspaced with sustentacular cells and stem cells (Figure 20).

Olfactory neurons possess an astonishing capacity for being recycled very rapidly. It has been calculated that the turnover period for these cells is about two weeks. This exceptional power of regeneration is quite amazing, as the neurons of our brain have almost totally lost such capacity. In fact, in adulthood our brain loses a number of neurons every day, which are not replaced. The peculiar renewal of olfactory neurons is due to the presence of primitive and pluripotent stem cells, able to transform into more than 1000 types of different olfactory neurons. Moreover, scientists have managed to generate normal mice by inserting the nuclei of mature olfactory neurons into egg cells. As may be imagined, these nasal stem cells have recently received much attention as potential candidates to replace the embryonic stem cells currently used for medical applications.

The olfactory neurons, similar to those which are part of our brain, are made of a cellular body with two long tails. The first extends across the whole width of the epithelium and reaches the external

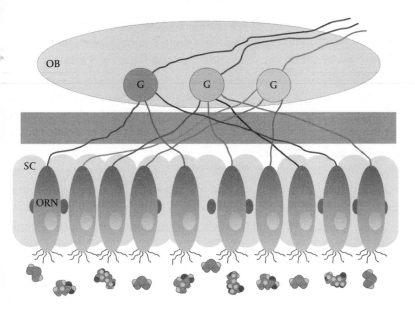

Figure 20. The olfactory epithelium is located on the upper region of the nasal cavities. Olfactory neurons (ORN) sit across the olfactory mucosa flanked by sustentacular cells (SC) and other cells and send their endings into the external environment. They terminate with tufts of cilia, which expand the surface exposed to odours. At the other end of the neurons, long axons cross the ethmoid (a perforated bone of the skull) and reach the olfactory bulbs (OB), where each glomerulus (G) collects the signals from all the ORs responding to the same odour and sends information to the central areas of the brain.

environment. Here it opens to the outside environment in a cluster of cilia like the petals of a flower, ready to catch the molecules of odorants. In fact, the main function of these cilia is to increase the surface of contact between the neuron and the external environment.

It is right on the surface of these cilia that the receptor proteins are located, embedded in the cellular membrane. These proteins, sitting at the entrance of our nervous system, like sentinels at the gates of a walled city, are responsible for checking the identity of each coming molecule and sending their messages to the interior of the cell. In this process, different chemical structures are recognized and discriminated, and an electrical signal is generated by the cell. All this relies on a correct fitting of the odorant molecule with specific cavities present in the structure of receptor proteins.

The other process of the neuron, called the axon, crosses a perforated bone—the ethmoid—and reaches the brain, where it establishes connections—synapses—with a second layer of neurons. This hub for the collection and distribution of olfactory signals consists of two olfactory bulbs, situated just below the front of the brain, one on the right and the other on the left. From the olfactory bulbs the signals are sent to the central areas of the brain where they are further processed, evaluated, and compared with the data stored in the memory finally to generate a conscious perception, a verbal description, or a behavioural response.

Biochemical research in the sense of smell started at the end of the 1970s, but we can regard the identification of olfactory receptors of vertebrates in 1991 as the major landmark in the field.[27] This discovery won Linda Buck and Richard Axel the Nobel Prize for Medicine or Physiology in 2004, providing an enormous stimulus to research and boosting the interest of scientists in olfaction.

Now, thanks to genome sequencing, a large amount of information is available on the number and identity of olfactory receptors. However, only for a handful of them have the corresponding odorants been identified. Therefore, although the path has been paved, we are still far from cracking the olfactory code. Even less do we know about the pathways which the olfactory messages have to follow across a complex network of neurons in the brain eventually to produce a behavioural response, an emotion, or a verbal description.

THE FIRST ATTEMPTS OF BIOCHEMISTRY

It was not until the end of the seventies that scientists interested in olfaction started talking about receptor proteins and the strategies to identify them. A few isolated attempts had been made, with dubious and non-reproducible results. Olfaction was still such an unexplored field that experienced biochemists did not dare risk their time and their budgets on research based on vague hypotheses and without reasonable prospects of success. Research is a highly competitive game and requires substantial funding, which in the end should be accounted for with solid scientific results. Therefore, investing in a completely new field requires either courage or the blind enthusiasm of a young scientist.

The very existence of olfactory receptors as proteins was still questioned by some scientists in the field. In fact, among several theories and models that were later proved inconsistent in the light of experimental results, some scientists argued that olfactory receptors could be nothing more than lipid molecules, while others went so far as to deny a direct interaction of odours with biochemical structures in the nose. However, the idea that receptor proteins are the biochemical elements responsible for detecting the volatile molecules of odorants became more and more acceptable and convincing as knowledge about other types of receptors, such as those involved in the transmission of nervous signals, was rapidly accumulating.

Still, the task was far from easy. Even now, after the sequencing of the human genome and of those of many other species has provided us with a wealth of information, searching for a novel family of genes without any specific information would be a major problem. It would be like searching through the pages of a book where hundreds of thousands of gene sequences are reported, without any hint of how to recognize those we are looking for.

If at the present state of technology the discovery of a new family of genes still presents serious difficulties, 40 years ago the task was much more challenging, almost desperate. The only tools available at that time were those of biochemistry, which involved direct studies at the protein level. The approach usually originates from a physiological recording or a behavioural observation—for example smelling an odour—to search for the protein responsible for that phenomenon.

The quest for a new receptor usually involves several steps. First, you need a reliable, rapid, and inexpensive (in terms of the sample used) test to measure the activity, and therefore the presence, of the receptor under study. For example, a drug or generally a chemical substance triggers an observable and measurable effect in an individual or an isolated organ. The next step is to study how small variations in the structure of the molecule affect the quality or the intensity of the produced signal.

Generally, receptor proteins sit across the membrane of the cell, for instance a neuron cell, and send information about the presence of foreign molecules inside the cell. The aim is to isolate such proteins and study their characteristics. In several cases, the same molecule recognized by the receptor protein has been used to fish out the receptor, like a bait, which gets bound to such a substance and can be isolated from all other chemicals and proteins present in the mixture.[28]

These techniques are similar to those that had been applied much earlier to the study of enzymes. An enzyme is also a protein which, like a receptor, first binds a chemical compound, but then performs a chemical reaction, modifying its structure and eventually releasing the product of its action (Figure 21). This cycle can be repeated many times by the same enzyme molecule, greatly amplifying the effect, in this case the chemical product. We can then understand that the accumulation of large quantities of products generated by minute amounts of enzyme make the study of enzymatic activity relatively easy. You only need to incubate your preparation with the enzyme's substrate and let the reaction run for a time long enough to reach a concentration of the product which can be easily measured.

Once a simple and reliable method for detecting the presence of our enzyme has been set up, its isolation and purification from a crude biological extract would involve some fractionation procedures. In each case, we end up with a series of test tubes, which can be assayed for the presence of the desired enzyme. Then, the tubes containing the enzyme can be combined and subjected to a second purification step, and so on until our enzyme has been purified from all the other proteins.

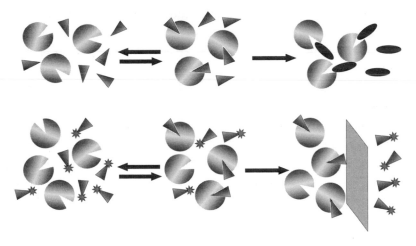

Figure 21. Differences between enzymes and receptors. The upper part of the figure shows molecules of an enzyme (spheres) which reversibly bind a substrate (triangles) and turn it into a product (ellipses). The activity of an enzyme can easily be measured by monitoring the amount of product. In the case of a receptor (lower part of the figure) no reaction occurs. The binding activity can only be measured after separating the proteins from the free ligand. If the ligand is 'labelled', for example with a radioactive tag, we can simply compare the radioactivity in the two fractions to evaluate the affinity of the ligand with the receptor.

In the case of a receptor protein, the procedure can be similar, but it is more difficult to set up a method to detect the presence of a protein which binds the ligand but does not perform any reaction (Figure 21). Several protocols have been developed, which however are relatively complex and not very accurate. The classical approach, particularly when the ligand is a small organic molecule, makes use of radioactively labelled chemicals.

The solution containing a receptor to be analysed is mixed with the radioactive ligand until equilibrium is reached. At this point, part of the ligand is bound to the receptor protein and part is still free in the solution. The ratio between the two concentrations is related to the affinity of the receptor towards the ligand. How then to measure the concentrations of free and bound ligands? We need to separate the free ligand from its complex with the protein, a task that can be accomplished by different techniques used to fractionate chemicals according to their molecular masses. Therefore, the free ligand will be that found in the low molecular weight fraction and that which is bound to the receptor protein will migrate with the protein and can be found in the high molecular weight fractions.

Such were the methods employed at the end of the seventies to identify and isolate receptor proteins. At that time most of the studies had been focused on receptors for hormones, particularly steroids, and for neurotransmitters, such as the acetylcholine receptor or the β-adrenergic receptors.

With such a background of expertise, assuming that olfactory receptors would be proteins, like receptors for other molecules, we might wonder why biochemists appeared to distance themselves from this new and exciting area. Scientists generally are always on the lookout for new ideas and fields to investigate. Indeed, it is the dream of all those involved in scientific research to be the first to discover a new molecule, a physical law, or a particular physiological effect.

The main reason was probably the high risk involved in a completely new research area. There is constant pressure to produce published material, which would support future request of research funds and often there is simply no room left for original and entirely new projects. When applying for funds, furthermore, it is important to present a project based on established ground, to support the presumption that in the end concrete results will be obtained. A new project, based only on ideas, has little chance of being funded.

Another aspect which discouraged biochemists from investigating the olfactory world was its extreme complexity. Even in the light of the scarce knowledge available at that time, experienced scientists could perceive that the olfactory system must necessarily be based on a large number of receptors, unlike the simple code of colour vision. Moreover, because of their great number, the quantity of each single type of receptor protein present in the nose was expected to be extremely small. Therefore, biochemists who wanted to do research in olfaction had to face the unique problem of identifying a single specific protein present in tiny amounts (below the detection limit of instruments and techniques in use at that time) and separate it from hundreds of structurally similar proteins. The task was far beyond the means then available.

This being the biochemical research position in the late 1970s, we can understand the lack of interest in a quest for the olfactory receptors at the protein level. Only a few attempts were reported, all of which were unsuccessful. In any case, results obtained in single laboratories were not reproduced by other scientists and soon lost perceived reliability and credibility.

Now, in the light of recent knowledge acquired through the more powerful techniques of molecular biology, the task that scientists had attempted appears quite impossible. In fact, membrane-bound receptors are present as a monomolecular layer on the surface of the cell; therefore their amount is far lower than those of proteins contained inside the bulk of the cell. Such a serious handicap was also aggravated by the presence of a large number of receptor proteins with similar properties within the same piece of tissue.

Besides all these problems, which are typical of receptors and in particular of olfactory receptors, the task of purifying a protein is far from easy. Generally speaking, to isolate a protein we have to separate it from a great number of other proteins, all similar in their chemical properties. In fact, all proteins are long chains of the same 20 building blocks, the amino acids, linked to one another in a linear arrangement, typical and constant for each protein. Therefore, proteins differ in their length and in the proportions and arrangement of the 20 amino acids. As some amino acids carry charges, proteins can be differentiated from one another on their total charge, as well as on their size. These are the two main characteristics that we can use to fractionate a protein mixture. If the protein of interest is one of the mixture's minor components, its purification may pose difficult problems and involve several fractionating steps.

The second major difficulty relating to olfactory receptors was the absence of a reliable functional test to detect their presence. At the end of the seventies there were not many examples of receptors that had been purified. In any case, for each of them a specific ligand was known, which could be used in a relatively simple and rapid assay to monitor the presence of the receptor protein during purification steps.

With olfactory receptors, on the other hand, the ligand constituted a major question mark. Any volatile molecule able to reach our olfactory mucosa is a potential odorant and a candidate ligand for olfactory receptors. However, not all odorants produce equally strong sensations. As we have already seen, some compounds are very weak odorants, while others are extremely potent. These latter include the familiar bell pepper-scented pyrazines, geosmin, responsible for the wet soil odour or androstenone endowed with the powerful pungent smell of urine.[29]

These molecules can stimulate our receptors at extremely low concentrations, therefore they could qualify as the strongest ligands, each one for its specific receptor. The same compounds should also reasonably be more specific than others and probably represent the baits of choice for hooking their corresponding receptors. Such was the idea behind the biochemical studies at the beginning of this search. It is easy to understand the usefulness, at that stage, of the large wealth of information accumulated during the preceding decades on odour descriptions and olfactory thresholds.

However the main difficulty still remained, how to detect and isolate a single type of receptor among hundreds of proteins of the same class. In this respect, the amount that could be isolated even from a large animal, such as a cow or a pig, became the major point of concern. A simple calculation, based on the area of the olfactory region, even assuming that such an area would be completely covered with receptor proteins, returns a value of about 100 micrograms as the maximum amount of olfactory receptors that could be packed in the nose of a single animal of large size. This figure should then be further divided by the number of possible receptor types (more than 300 in humans and close to 1000 in other animals, as we subsequently learned) to get an estimate for each type of receptor. Therefore, any experienced biochemist knew that this was an impossible task and would not even think of investing in such dubious research.

A QUEST FOR OLFACTORY RECEPTORS

At that time I was young and inexperienced, particularly in biology (having been educated in organic chemistry), and without funds. I had nothing to risk and my ignorance prevented me from seeing the absurdity of the mad adventure I was going to undertake. If I had paused and considered all the risks involved; if I had asked advice from experts in the field or made a more detailed study before embarking on a search through unexplored land; if I had been older, more mature, and experienced, I would probably never have known the excitements and joyful moments of the past 35 years, accompanied though they were by frustrations, disappointments, and failure, all common ingredients in the everyday life of a scientist.

Following my impulse I began my quest for olfactory receptors. The first step involved finding a promising ligand to use as bait for the receptor protein. Such a chemical would be selected from the most powerful odorants, based on the theory that a strong odour would be associated with close interaction between the ligand and the receptor protein.

My first choice fell on androstenone, the urine-smelling steroid, which we have already discussed more than once. This molecule is endowed with an exceptionally low olfactory threshold, and besides, it represents one of the clearest examples of specific anosmia, being perceived by only half of the human population. This fact, in particular, reveals the presence of a very specific receptor behind the detection of this unique odorant. Androstenone is also a well studied sex pheromone for pigs, a fact supporting the presence of a dedicated receptor in this species and probably in other mammals.

Therefore this steroid seemed to be the ideal choice and likely to produce the expected results.

In fact this did prove to be the winning card, but only much later for Leslie Vosshall, Hiroaki Matsunami, and their co-workers, who in 2007, exactly 30 years after specific anosmia to androstenone was first reported,[30] were able to identify the olfactory receptor whose malfunction generates such phenomenon.[31]

At the end of the 1970s, however, androstenone turned out to be too hard a challenge for the available analytical tools and techniques, but nevertheless became the main character in the events which took place in the following years when scientific research and human ambition were the ingredients of a sort of detective story with twists and turns that we could not have imagined at that time.

As the amount of receptor likely to be present in the olfactory mucosa of even a large animal, like a cow or a pig, was expected to be extremely small, it was necessary to use radioactively labelled ligands, in order to meet the sensitivity required in these experiments. Using such radioactive probes, we would be able to detect quantities down to the order of a picogram (one millionth of a millionth of a gram). Therefore, the first step was to synthesize a *precursor* that could be made radioactive in specialized labs. The precursor was the molecule of androstenone itself which, being endowed with a carbon-carbon double bond, can easily be turned into its analogue androstanone, very similar in structure and odour, and labelled with two atoms of radioactive hydrogen.

As the compound was not commercially available at that time, I decided to synthesize it through a series of chemical reactions already described in the literature. Everything went smoothly until I reached the last step when androstenone itself was generated from an odourless compound. There were only a few milligrams in my preparation, but it was enough to fill all the rooms in my department with its repugnant smell. The foul and heavy stench of stale urine was lingering in the labs and reached the common room, where we used to meet for coffee, and stuck to the wool of our clothes—unsurprising because androstenone binds strongly with proteins (and wool is nothing other than protein) and everybody inevitably took home more than a few molecules—certainly enough to be detected and to place us in embarrassing situations. It is safe to assume that during this time, several visitors to my home were probably convinced that washing was not an activity routinely practised in my family.

After having the radioactive probe synthesized from our precursor, we finally began the search for the needle in the haystack. The haystack was a mixture of hundreds of different proteins extracted from the tissue lining the nasal cavities. We chose the cow and the pig as our model animals mainly because of their large size and the easy availability of biological material.

The work usually started with a visit to the local abattoir very early in the morning, where we would cut bovine and pig heads open, get the olfactory tissue, put it on ice and rush to the lab to extract the proteins. Once the extract was ready, this was incubated with a tiny amount of radioactive androstanone, allowing the probe to diffuse in the solution and bind with any suitable protein (hopefully the olfactory receptor), which would become labelled with its radioactivity. The mixture was then separated from the excess of ligand in order to measure the amount of androstanone associated with the protein. This step was carried out as rapidly as possible, to prevent detachment of the ligand from the protein.

Overall it appeared to be a straightforward procedure, except for the fact that the protein of interest was a negligible component in a very complex mixture. As a consequence, the amount of androstanone non-specifically bound to other proteins usually far exceeded that supposed to be specifically linked to our hypothetical receptor. One of the reasons was the fact that androstanone is a very hydrophobic molecule, with 19 carbon atoms and a single oxygen, this being the only atom in the molecule capable of interacting with water. Therefore, androstanone and similar compounds do not like remaining in water and would rather adhere to whatever else is around, like proteins and even the glass of test tubes. At this stage we started to suspect that our choice of androstanone probably was not the best.

The experiments were long and tedious, proceeding for the most part in the cold room, and the results were always poor. No clear evidence could be found that a protein specific for androstanone even existed in the extract. While we were struggling to get some vaguely reproducible results, a paper appeared in one of the leading journals of biochemistry, reporting the results we had been dreaming of obtaining, using the same protocol. Everything worked perfectly, the evidence was clear that there was a very specific protein that could bind androstenone selectively and reproducibly.

This of course proved that my idea was right, but that was only a meagre and bitter consolation in the face of the fact that I had missed a golden opportunity. It was clear that I had overlooked some practical but essential details and failed in my goal. In such situations, apart from the disappointment of not winning the game, there is a certain feeling of inadequacy. Many scientists cannot stand failure and such difficulties are often discouraging enough to convince them to abandon research. A good scientist is supposed to possess many other qualities, apart from being able to produce good science. Among them, there is certainly the capacity to withstand failure and to be able to lose and start again, like a card player or a boxer.

In science, to arrive second is like not arriving at all. Only the first over the finishing line gets the credit. However, quite often the credit goes not to the one who first generated the idea or produced the results, but the first who got the data published—not always the same person. Such strong competition to publish first sometimes leads to unethical behaviour. Although uncommon, there are cases where scientists, aware that another colleague has submitted to a journal results similar to the ones they are producing, attempt to delay the publication of the manuscript in order to publish their own first.

It can also happen that the pressure to publish is so high that manuscripts are submitted (and sometimes accepted) before enough results have been collected to warrant reliability or else with data that has undergone extreme *cosmetic* treatments.

My disappointment was certainly great, but at the same time this publication further stimulated my decision to carry on with this line of research. This was only the beginning and there were still many aspects to be investigated. The proof that such a receptor existed only represented the basis on which to plan all future work: the isolation of the protein, its characterization, how it was folded, and where the ligand would fit inside its complex structure.

Then, of course, there were all the questions regarding physiological aspects relating to such a receptor. How is the chemical signal translated into electric impulses and how are all the wires connected from the neurons of olfactory mucosa to the higher areas of the brain, which eventually lead to perception and behaviour?

The first thing to do was to reproduce in my lab clear and accurate results. At the time I was in touch with another colleague, Krishna Persaud, who was working at the University of Warwick, in England, in the lab of George Dodd, a scientist as mad as I was in pursuing this impossible search, and we all decided to concentrate our efforts in repeating the experiments on the androstenone receptor. We performed the same experiments and got the same poor and inconclusive results. Strangely enough, there was significant agreement between data collected in our two labs, but not with those that had been published.

After struggling for several months and persistently (and frustratingly) collecting unsuccessful data, I decided to abandon androstenone and chose a more hydrophilic odorant for my experiments.

To the present day, those results that were published have not been reproduced by other scientists and appear to be the product of inaccurate experiments.

AN UNEXPECTED DISCOVERY

Sitting amongst the ashes of the great androstenone failure, I looked for a new probe, which would not present the inconvenient hydrophobicity of steroids. The choice fell on another extremely potent odorant, which we have met, the bell pepper smell, 2-isobutyl-3-methoxypyrazine. This molecule has an odour as strong as, or perhaps stronger than, androstenone, but has the advantage of being much more hydrophilic. It is still an oily compound that generally speaking would be classified as *insoluble*. However, at least one gram can be dissolved in a litre of water—far more than we needed for our biochemical experiments. As in the case of androstenone, the exceptionally powerful smell of this pyrazine suggested strong and specific interaction with olfactory receptors.

I started again with the new odorant and synthesized the pepper-smelling pyrazine to have it radioactively labelled. This time my lab was filled with a fresher and more pleasant smell, a great and welcome change from the nasty, pungent stench of androstenone. But, however pleasant and natural this odour was, because of its exceptional strength the entire department was constantly immersed in a bell pepper atmosphere. Anything stored in the fridge where a sample of the pyrazine had been kept soon acquired its recognizable character. It seemed that such a smell had the power of reproducing itself, like a virus, and was continually invading new areas. A simple calculation, based on its very low olfactory threshold, will easily demonstrate that only a few milligrams, just a tiny droplet, could be well perceived even if dispersed in a volume as vast as a large building.

Working with the pyrazine was a different story. The results came out easily, and were clear and reproducible. In just a couple of weeks I had good data to show to my colleagues in Warwick and obviously felt very proud. However, I noticed that something was not quite right, but could not understand why. It was a sort of hunch that everything looked too simple. All the difficulties I had foreseen seemed to have dissolved. I repeated the experiments several times and the reproducibility was excellent. Here I had a nice saturation curve, with a very low background (the main difficulty I had experienced with androstanone) and plenty of material to work with. In fact, this was the main problem, there was too much of our putative receptor.

I made a rapid calculation, considering the number of olfactory neurons present in each square centimetre of the mucosa and the total area of the ciliary membrane. But, even assuming that the entire membrane surface was covered with receptor proteins, the amount I could expect for each olfactory was about 10,000 times less than that which I was measuring.

There was only one likely explanation, although hard to accept: my protein was not an olfactory receptor. It could have been any protein that, by coincidence, had some affinity for the molecule I was using as a probe. Such a rational explanation was not easy to accept emotionally. I could not let all the euphoria of the previous weeks be quenched by the obvious data spread in front of me. Was it really the case that we had been wasting time with a protein that had nothing to do with olfaction?

To address these issues, a few simple binding experiments performed with extracts from other tissues would tell us if the measured activity was only specifically found in the nose. This is a common procedure when we want to relate biochemical data to physiology and investigate the function of a given protein. We therefore subjected extracts from liver, brain, spleen, lungs, and several other tissues to our analysis. All of them proved to be negative so that we could confidently conclude that the property of binding our pyrazine compound was specific to nasal tissue. This certainly gave us confidence that whatever we were measuring was likely to be related to olfaction. All was not lost. We had discovered something that might prove interesting, although the results did not fit our model.

We then decided to put this as yet unknown protein to another test. When identifying a new receptor, one of the most convincing arguments that you are looking at the right type of receptor is to show that your *in vitro* preparation recognizes different ligands with the same specificity as the natural receptor does *in vivo*. The physiological data to compare with the biochemical results were in this case the olfactory properties of volatile molecules, such as odour descriptions and olfactory thresholds. It was the right time therefore to make good use of the wealth of data accumulated by chemists during the past decades. In this respect, pyrazines and similar compounds had been the object of active research by scientists interested in olfaction, because they represent an important class of food odorants.

Therefore, we chose a series of compounds of similar chemical structure, but which differed in their smell. The pyrazines also proved particularly suitable for this kind of investigation because of their simple, relatively rigid structures, allowing for small modifications to be introduced on the molecules in order to study their effects on smell. A large number of studies had addressed such questions, providing a wealth of information and establishing clear, solid relationships between structure and smell.

In particular, as we have already observed, a change in the length of the hydrocarbon chain of 2-alkyl-3-methoxypyrazines produces a drastic effect on its smell. While derivatives with methyl or ethyl groups are *nutty* and *roasted*, the presence of a chain of three or more carbons turns the compound into a green vegetable odorant, while dramatically increasing its strength.

When we tested the *green* and the *nutty* pyrazines in our assay, we observed completely different behaviour between the two classes. The results provided great excitement, as they turned out just as we hoped: this puzzling protein, whatever it was, recognized the different pyrazines owing to their smell. Then we went a step further and tested, in the same kind of experiment, another series of chemical compounds, alkyl substituted thiazoles. In this instance as well, derivatives with short chains (one or two carbons) smell *nutty* and *burnt*, while increasing the chain length produces compounds with a *green* character. Again, the binding experiments matched the olfactory data.

Such results met with great excitement. We submitted a manuscript to the *Biochemical Journal* and it was readily accepted and published in January 1982.[32] Although we were pleased with such unexpected results, the presence of this protein in the olfactory organ, instead of contributing to clarifying the process of odour perception, merely added confusion to the narrative. Here we had a protein, specifically expressed in the nose, able to recognize volatile molecules according to their smell; but it could not be one of the receptors we were seeking, because there was too much of it.

It was a completely unexpected discovery, certainly interesting, but disappointing at the same time. However, it proved to be the beginning of an exciting story which is far from being completely developed.

This discovery, as so often in science, was made by accident. It is common that the results of scientific research are different from those expected, sometimes disproving the predictions at the basis of the project. In such cases, the first impulse is to discard everything and start with a new proposition. But this is the very situation in which we recognize the good scientist. Instead of sticking to his original idea, such a scientist critically examines the results produced and often finds something even more striking and unexpected. A dirty chunk of rock can conceal a gold nugget. The history of science is replete with such anecdotes.

For example, we could cite the case of ionic liquids, discovered, perhaps we ought to say recognized, in relatively recent times. In fact, ever since the establishment of chemistry as a science, researchers had been struggling with chemicals that refused to crystallize, and with disappointment they often abandoned the project. Ionic compounds are called salts, just like table salt or many different minerals which produce hard, clear crystals. These contain two parts, namely the negatively charged anion and the positively charged cation. These two parts are held together by electrostatic interactions and usually are assembled into the well ordered structures which we call crystals.

This is the rule, but there are exceptions, the so-called ionic liquids.[33] When the structure of the anion and that of the cation are rather complex and cannot be neatly arranged in a simple pattern, the product refuses to crystallize. Many an organic chemist is aware of the frustration and disappointment accompanying the struggle to crystallize a reluctant new compound, in the end throwing everything down the sink. How many times were ionic crystals produced, only to go unrecognized?

It was fortunate that we did not discard this protein. Although we could not make much sense of the discovery at the time, intrigued by so many inconsistencies, we decided to investigate further. Odorant-binding proteins, as these new proteins became known later, have been the object of intense research, which has provided a vast amount of information in the last three decades, while opening new perspectives and raising unanswered questions in the field of olfaction.

8

ODORANT-BINDING PROTEINS

A Family of Versatile Molecules

THE BOVINE ODORANT-BINDING PROTEIN

The discovery of the bovine olfactory protein, which was later called OBP (odorant-binding protein) proved to be more interesting and full of consequences than we had imagined. It marked a turning point in olfactory research, as more scientists regarded a biochemical investigation of the olfactory system as something feasible and chose to dedicate more attention and resources to such aspects.

First step: purification

The next step was to get this protein purified, in order to obtain detailed information on its structure and mode of ligand binding, and perhaps formulate hypotheses on its physiological function. At that time Krishna Persaud, who had shared with me the frustration of the adventure with androstenone, and with whom I went on to establish a long lasting collaboration and a deep friendship, had completed his PhD in George Dodd's lab at Warwick and then joined me in Pisa to help in the characterization of this new intriguing protein. Almost at the same time, another group in Italy, led by

Andrea Cavaggioni at the University of Parma, became interested in the project and we decided to join forces to obtain a sample of the protein pure enough to complete its characterization.

In the absence of any structural information, we needed to follow this protein through the subsequent purification steps by tracing the bound radioactive pyrazine (that odorant with the smell of bell peppers). It was fortunate that the interaction between the labelled pyrazine and the OBP was strong enough to survive all the column elution steps, so that by measuring the radioactivity in the eluted fractions we could easily spot those containing our protein. In the end the purification process proved easier than expected, thanks to the abundance of this protein and to its favourable chemical properties.

Nevertheless, the task took about three years. We had been working at a snail's pace, and wasted time in minor experiments, which did not in the end add any substantial information. The real reason was perhaps the fact that we had not yet appreciated the interest that such a discovery would arouse. After the initial excitement, the certainty that we were still far from identifying the olfactory receptors subdued our enthusiasm for further work and we lapsed into a sort of routine, that at times seemed boring and aimless. We did not, however, feel any pressure or competition from any other research group, as we still were the only ones taking such an approach. But soon we would learn that the situation was evolving very fast.

Competition stimulates research

By the time we had our data almost ready for publication, we heard rumours about another group in the United States, who had already prepared a purified sample of the same protein. Solomon Snyder, at Johns Hopkins, Baltimore, had not long before attached his name to the discovery of opioid receptors, for which he had received the prestigious Lasker Prize. When we learned that we were competing with such an experienced and powerful research group, the first impulse was to think of abandoning our research for more pleasant and mundane activities. The much wiser course, however, was to finish our experiments without further delay and accelerate the publication of our data. In the end, the two papers were published at about the same time, ours in the European Journal of Biochemistry,[34] Snyder's in the Proceedings of the National Academy of Sciences.[35]

An increased interest followed the publication of these papers and the entire field of olfaction was shaken from the lethargy of the previous period. Researchers had been waiting for a new, more direct approach to the discovery of the olfactory code and the molecular mechanisms responsible for perception of odours. Now they had a new tool, proteins which could recognize odorant molecules. Were they acting as some sort of olfactory receptors? Of course, being soluble proteins, they should be either contained inside the cellular fluid or floating in the extracellular space. Instead, to convey a message from the external world to the interior of the olfactory neuron, we need proteins sitting across the cell membrane, with part of the molecule exposed to the environment to detect chemical signals and part inside to activate the enzyme machinery of the neuron.

On the other hand, the soluble nature of these binding proteins and their secretion close to the membrane of the olfactory neurons, where the true olfactory receptors were assumed to be located, strongly recalled the soluble proteins of bacteria. These are found in the periplasmic space between the inner membrane and the external wall, and are able to bind sugars and amino acids. Such bacterial proteins had been classified as *receptors*, a type of soluble receptors, and our olfactory proteins could perhaps perform a similar function.

Whatever their name, these were the only proteins so far shown to be interacting with odorants and several scientists decided it was worth spending energy and resources on studying the properties and the function of such new actors on the scene of olfaction.

The high confidence in the published results was also due to the fact that two different groups, ours and Snyder's, who had been working independently and without exchanging information, had reached the same conclusions. In fact, the best evidence for a new discovery is the reproduction of the results by other scientists. This was the first time that a reliable and reproducible result was obtained in the biochemistry of olfaction. Another fact that increased the interest and the reliability of our discovery was the report, around the same period, of another olfactory protein, this time in insects.

Richard Vogt and Lynn Riddiford had identified an abundant soluble protein in the antennae of a giant moth, *Antheraea polyphemus*, able to bind the sex pheromone of this species and therefore named PBP (pheromone-binding protein).[36] As we shall describe in more detail, this protein, although different in structure, presented all the features of the insect equivalent of our mammalian OBP and could possibly be involved in similar functions, whatever those may be.

In order to gather wider information on these proteins, we started purifying OBPs from different animal species and measuring their affinities to several compounds endowed with strong odours. The weak point we had been worrying about since the beginning of this story was the fact that all the olfactory data had been recorded with humans, while the biochemistry had only been studied in other mammals. Although we could reasonably assume that strong similarities existed in the olfactory systems across mammals, it was nevertheless desirable to investigate the properties of OBPs in humans.

Several groups became interested in searching for a human OBP, yet all attempts failed. Later, after genome sequencing, the human OBP was finally detected, first as a gene and then as a protein, which proved to be produced only at extremely low levels—a situation completely different from those of the cow or the pig, from whose nasal tissues large amounts could be easily obtained, up to several milligrams from a single animal.

Second step: amino acid sequence

As soon as the bovine OBP was purified, we started investigating its structure. The first step was to determine its amino acid sequence. Although this was a small protein, with about 150 amino acids, at that time, before the tools of molecular biology had found wide application in the biochemical labs, such a task was far from simple. It took us almost one full year to complete the sequencing of the bovine OBP, performed with the traditional biochemical approach. Certainly this protein and the pig OBP, whose structure came next, were among the last to be sequenced in this way. In fact, soon after, the methods of molecular biology, much faster and cheaper, completely replaced the traditional system.

It is still worth describing the biochemical method for sequencing a protein, which was cumbersome compared to current DNA sequencing procedures and is now obsolete. In practice, we subject a sample of the protein to a series of reactions that have the effect of chopping one amino acid residue at a time. The reaction starts at the amino terminal (by convention regarded as the beginning of the protein chain) and can be efficiently repeated for 20–30 residues, 40 at most. The reaction sequence was devised by Pehr Edman in 1956 and remains unchanged, testifying to how brilliant his discovery was. This procedure, which at the beginning was done manually and required huge amounts of purified protein, was soon performed by dedicated machines, assemblies of pumps and valves, called amino acid sequencers, which could complete the work overnight.

But even a small protein, like an OBP, is much longer than the 20–30 amino acid that could be directly sequenced. Therefore, in order to obtain the full length, the protein has to be cut into fragments, by using specific enzymes, which cut bonds at defined positions. The fragments are then separated using chromatographic techniques and individually sequenced. Then, in order to assemble all the fragments correctly, we still need to process another sample of the protein with a different enzyme, which cuts this protein at different positions. Again the fragments have to be separated and sequenced. In this way, we can get overlapping regions that enable us to reconstruct the complete sequence of the protein, like a jigsaw puzzle. This procedure is long and requires relatively large amounts of the protein.

Now, using the techniques of molecular biology, we can obtain the complete sequence of a protein in a few days. The method currently employed only requires a short segment to be sequenced, about 6–10 amino acids, preferably the first in the sequence. This information allows us to design and synthesize an oligonucleotide, a short fragment of DNA, containing the sequence of bases coding for that particular sequence of amino acids. Each amino acid is coded by a triplet of bases, therefore the gene encoding 8 amino acids is only 24 bases long, and was easy to synthesize even at the early stages of these technologies (now we can routinely synthesize an entire gene for a small protein, several hundreds of bases long). This oligonucleotide is then used to fish out and amplify the gene encoding the protein under study, using a technique that has rapidly become one of the most common tools in all labs of molecular biology. It is called PCR (polymerase chain reaction) and was devised by Kary Banks Mullis in 1983, an achievement that won him the Nobel Prize for Chemistry 10 years later.

The idea of replicating DNA, thus multiplying its amount, is certainly not new. In fact the most interesting property of DNA is its ability to make copies of itself: it is the reaction that makes life possible and controls the transmission of hereditary characteristics to the next generation, as Watson and Crick suggested in their seminal paper in *Nature* describing the structure of DNA. Such a characteristic is based on the complementarity of the four bases (A: adenine, G: guanine, C: cytosine, T: thymine) which constitute the molecule of DNA. They can establish relatively strong and specific interactions using *hydrogen bonds* where hydrogen atoms linked to oxygen or nitrogen can act as bridges between two of these atoms. The interesting aspect is that hydrogen bonds are strong enough to establish specific interactions, but at the same time weak enough to be easily broken, while keeping the structure of the molecules, which is held together by much stronger covalent bonds, that are unaffected.

During the process of replication, the single nucleotides bind to the DNA molecule, which acts as a template through specific hydrogen bonds (adenine couples with thymine, and guanine with cytosine), while an enzyme joins these new building blocks to one another like a sewing machine. When the second chain of DNA, complementary to the first one, is completed, the two segments can be separated merely by heating the mixture and the process can start again. During the PCR this is done by first increasing the temperature to about 95°C to separate the two strands of DNA, then lowering it to around 50°C to allow pairing with the specific 'primers', then increasing it again to 72°C during the 'elongation step' when nucleotides are added one by one to the growing chain along a pattern complementary to that of the template. Typically a PCR protocol involves 30–35 cycles of this type. At every cycle, the number of DNA molecules doubles: we can easily calculate that a single molecule can yield at the end of the reaction, at least in theory, one billion identical DNA chains.

The weak point of this reaction is the high temperature required to separate the two chains of DNA. Such harsh treatment, repeated several times, can quickly affect the performance of the enzyme which, being a protein, can easily be denatured at such temperatures. In fact, the key element that made PCR a practical tool was the availability of enzymes (polymerases in this case) which withstand high temperatures.

The isolation of such enzymes was the consequence of a much wider discovery in biology, that of organisms capable of living in extremely harsh environments. Often they are micro-organisms, appropriately named *extremophiles*, which can survive in conditions prohibitive to all other living organisms. Some of them live near hot springs, where very high temperatures can be reached. In the depth of oceans, in some of these hot springs temperatures up to 120°C have been recorded. Other micro-organisms can survive in extreme cold or else prefer high concentrations of sodium chloride, such as those found in salt evaporation ponds. Surprisingly enough, some forms of life can only survive in such extreme conditions and suffer when brought into normal environments.

The discovery of PCR represented a milestone in science and a powerful tool for research, allowing for the analysis of microscopic samples of DNA, which so far had escaped any investigation and lowered the detection limit by millions or even billions of times.

We can compare the improvement brought by PCR to the progress made by the microscope or the telescope, which also enlarged scientists' field of observation. This technique has made possible the analysis of extremely small samples of tissue and thanks to such unprecedented sensitivity has found applications in forensic analysis, where the sample of a fingerprint or a hair can reveal the identity of the owner.

After a gene has been amplified, it can be sequenced in a much quicker and easier way than the protein, requiring only hours rather than months. The sequence of the gene can be translated revealing the amino acid arrangement along the whole length of the protein. The techniques of molecular biology have made possible the sequencing of a great number of OBPs, even when the amount of the protein which could be purified from natural sources, as in the case of insect OBPs, was too tiny for traditional biochemical methods to be applied.

Third step: three-dimensional structure

Another great tool provided by molecular biology is the possibility of synthesizing large amounts of a given protein by infecting a bacterium or another organism with the gene encoding the protein of interest and letting the bacterium produce the protein. The availability of OBPs on the scale of milligrams has allowed the preparation of crystals, which have been utilized in X-ray diffraction spectrometry for resolving their three-dimensional structures.

Thanks to the rapid development of these techniques, a great deal of detailed information is currently available on the structure of OBPs, on their ligand-binding pocket and their mode of interaction with odours and pheromones. Literally thousands of OBP sequences, both in vertebrates and in insects, are currently available and for several of them the three-dimensional structure has been solved. Moreover, thanks to the large amount of information available and the improved bioinformatic tools, we can model a new protein simply on the basis of its amino acid sequence, provided there is enough similarity with other members of the same family with known structure. The structure of a protein is of fundamental importance, because it is strictly linked to its physiological function and often the first hint of the role of a new protein is suggested by its three-dimensional folding.

At this stage we should perhaps move away from the history of odours and olfactory proteins and spend some time describing the structural elements of proteins and their relationships to function.

The shapes of proteins

We can first imagine a new-born protein (when it is synthesized by the cell machinery) as a long string, a chain of amino acids linked to each other by covalent bonds between the carboxyl group of each unit with the amino group of the next, like a number of people holding hands in a linear human chain. Each amino acid, beside the amine and the acid groups, bears a side chain, that could be as simple as a hydrogen atom or a small hydrocarbon chain, but could also contain other functional groups, including a second amine group or a second carboxyl group. There are 20 different amino acids making up all the variety of proteins and most of them are generally present in each protein. What distinguishes one protein from another are the different relative amounts of the 20 amino acids, but, more importantly, the arrangement of these different building blocks along the chain, a sequence absolutely unique for each protein. This sequence is encoded in the DNA and determines the three-dimensional structure which in turn is responsible for the physiological function of the protein.

Once synthesized, the long thread folds into a specific and unique shape, although at first glance a protein might look just like a randomly coiled string. Interactions between the functional groups present in the protein guide the folding first into small domains that are further arranged in the final three-dimensional structure. The most common domains are helices and pleated sheets: these are structural elements that are connected by short more flexible segments and assembled into the final shape of the protein. These domains are relatively small and usually involve 10–20 amino acids. Hydrogen bonds perform a basic role in stabilizing both helices and pleated sheets, and also contribute to holding these domains together, along with stronger connections, such as those between opposite charges or weaker hydrophobic interactions. As proteins are usually in an aqueous environment, the best arrangements of the amino acid chain are those that keep hydrophobic residues inside the core of the protein, while charged or hydrophylic groups are on the surface.

Solving the structure of a protein

Determining the three-dimensional structure of a protein is far from easy even today, and for some classes of proteins highly challenging. We cannot observe the shape of a single molecule of a protein with a microscope, however powerful. The first limit is set by the wavelength of the light we use. Visible light comprises wavelengths between about 400 and 800 nm (nm indicates a nanometre, equivalent to one millionth of a millimetre). Any object smaller that about 200 nm would thus appear completely blurred and impossible to identify.

A typical protein of small to medium size can be imagined as a little sphere with a diameter of around 3 nm, 100 times smaller than the limit set by light. Using a beam of electrons we can go down the scale, in fact electron microscopes have been used to obtain images of large pieces of DNA and even of proteins, which however appear like blurred dots without any possibility of spotting the positions of single atoms.

But to see the arrangement of single atoms in a protein we need better resolution still. A single atom, simplified as a small round object, would have a diameter of 0.1–0.2 nm and this is the resolution required for a protein structure. Instead of observing the atoms directly, we can calculate their positions from diffraction patterns originated by X-rays that interact with a crystal of the protein. This technique is not easy to explain and is also far from being straightforward in practice.

The first step involves growing crystals of a protein, a task requiring a large number of trials and whose success cannot be guaranteed. The second challenge then is to obtain a good diffraction pattern from the crystals and finally the data must be interpreted in terms of spatial coordinates. Of course, we have to assume that the structure of the protein in the crystal reproduces the natural folding of the protein when it is freely swimming in solution.

This assumption might seem far-fetched and unrealistic, but in fact it is very reasonable and can be adopted with a high degree of confidence. The reason for this is that crystals of proteins are quite different from the crystals we are familiar with in our everyday life, such as those of quartz, sugar, or table salt. In fact, when the molecules of proteins arrange themselves in ordered rows and columns, they take with them a large amount of water and retain the same environment as when they are in solution. A protein crystal, unlike the hard crystal of sugar or quartz, is extremely brittle and fragile and often can be destroyed even by gentle handling.

All the fine techniques behind the work to complete the structure of a protein lie beyond our area of interest at this moment and we will focus our attention on the results of such work and their consequences for the functions of our proteins.

As it is so complicated and difficult to get the three-dimensional structure of a protein, we could ask ourselves whether it is possible to obtain some hints from the amino acid sequence. We have said that, to a large extent, the final shape of a protein is in some way encoded in its sequence and most of the time the folding a protein assumes in solution corresponds to the most stable situation, that which can be reached spontaneously by the protein when in its suitable environment (pH, salts, ligands, etc.). Therefore, at least in theory, it should be possible to calculate the folding of a protein on the basis of its sequence. However, this is a very difficult task. But an easier approach can be adopted, based on the large amount of structural information currently available for many proteins, belonging to different families. We can assume, in fact, that proteins similar in their amino acid sequences might also share similar structures.

Therefore, as soon as sequence information is available, this is fed into a computer program, which searches for similar proteins in a very large data base. Finding members similar to our unknown protein, which have been studied for their structure or physiological function, can shed light on the properties of our new protein.

The structure of the first OBP

This was done with the bovine OBP as soon as partial sequence information was obtained. It was immediately clear that OBPs were part of a much larger family of proteins sharing not only amino acid sequence, but also a three-dimensional structure and some aspects of function. Several of these proteins had been already studied and the folding of a couple of them, such as retinol-binding protein, a carrier for retinol in the blood, and β-lactoglobulin, one of the main components of milk, had been resolved. They were later called *lipocalins* because they bind hydrophobic compounds (lipids and other molecules) and their shapes resemble a cup (calyx). The core of this cup is the site where the ligand (in the case of OBPs, the odorant) is located, a water-repellent region where the hydrophobic ligand can find a more suitable environment to sit, compared with the external aqueous medium.

Therefore, the next task was to obtain crystals and resolve the detailed structure of the bovine OBP, the first member of this family we had isolated. Again, this task took several years. The problem, in this case, was not the crystallization of the protein, which in fact yielded very nice, relatively large crystals with little effort, but the interpretation of the diffraction data. Being a lipocalin, we expected a structure similar to that of most proteins of this family, but the data did not fit such a model.

In the end, it became clear that, although the core of the bovine OBP is still in the shape of the classical cup (or basket), this protein was different because it was present as a dimer with the two units strongly hugging one another, exhibiting the so-called phenomenon of *domain swapping*. This means that one domain (a segment, in particular a helix) of one sub-unit was interacting with the core of the other unit and vice versa, like an arm protruding to embrace the other protein (Figure 22). This expedient obtains the effect of stabilizing the structure of the protein and making the dimer a unique compact body.

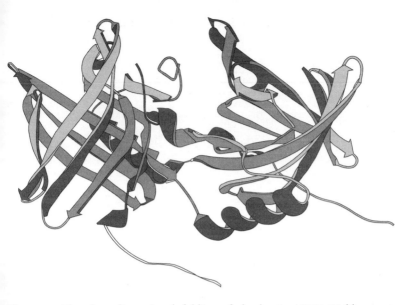

Figure 22. The three-dimensional folding of the bovine OBP. Unlike most lipocalins, this OBP exhibits the uncommon phenomenon of 'domain swapping' with the helix of one unit interacting with the core of the other unit. This mutual interaction firmly stabilizes the structure of the dimer.

The resolution of the structure of the first OBP, the one isolated from bovine nasal tissue, was the joint effort of an Italian group led by Hugo Monaco, of which I was part, and the American group of Mario Amzel, who was collaborating with Solomon Snyder. So, the same scientists who had been competing for the purification of the OBP found themselves collaborating on the same paper in which its structure was published. Competition can be stimulating in some cases, particularly at the beginning of a research project, as it generates new ideas and provides the necessary pressure to complete the work. Later on, however, this could result in a waste of funds and energies and does not benefit science but only personal ambition. In such cases, it is better to transform competition into more efficient and productive collaboration.

Competition again

However, as we were collaborating with the groups of Snyder and Amzel, another crystallography laboratory was working on the identical structure, curiously rehearsing the same events we went through during the purification of the bovine OBP. In 1996 the same journal (*Nature Structural Biology*) published two very similar papers reporting the structure of the same protein.[37] The group led by Christian Cambillau and Mariella Tegoni in Marseille had been working on the same project without being aware of each other's results and it was a happy coincidence that once again the results were independently confirmed. Soon after I established a long-lasting and still ongoing collaboration with Christian and Mariella, and more importantly a deep friendship.

The first product of this new collaboration was the structure of the pig OBP (Figure 23), a protein which exhibited a monomeric structure, more similar to those of other lipocalins. Again, we can see the typical basket image surrounding the binding pocket, where a molecule of a ligand has been represented. Why do these two similar proteins

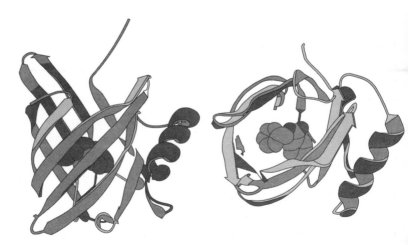

Figure 23. Two views of the three-dimensional structure of pig OBP complexed with a molecule of benzyl benzoate.

assume different shapes in the space? It is all down to the fact that the bovine OBP does not have any cysteine, a sulfur-containing amino acid capable of forming a covalent bond between two of these residues. These disulphide bonds play a major role in stabilizing the structure of a protein, because they connect two often distant amino acids, thus constraining the protein chain into a more compact conformation. While the pig OBP is stabilized by a disulphide bond, the bovine protein increases its stability by assembling two units in a single compact structure.

At this point perhaps I should make a couple of observations about the representations we have adopted. For a clearer visualization of the protein folding I have used a sort of cartoon, where only the main chain of the protein is evident and represented as a ribbon.

This way of visualizing the structure of a protein is a drastic simplification and is in a sense rather misleading. In fact, to this skeleton we should add all the side chains of the single amino acids with all the atoms packed together. If we perform this operation, which can be done with a single click on our computer model, we end up with an extremely compact structure with virtually no space left between atoms, except in the inner binding cavity.

This compact structure of OBPs and most of the proteins belonging to the family of lipocalins, is the reason behind their very high stability against any sort of degrading agent. We can literally boil these proteins for several minutes—a treatment that would irreversibly denature many other proteins—and then recover their full activity when returned to room temperature. Harsh organic solvents also have little effect and even proteases find it hard to cut through such compact folding.

Such exceptional refractivity to degradation is not surprising in proteins that are continuously exposed to the environment and to all the potential noxious compounds carried into the nose by the continuous flow of air. The same compact shape and stability bring two important consequences. The first is the high suitability of such proteins to biotechnological uses, such as in the fabrication of biosensors for odours. The second is that virtually all OBPs and many lipocalins are strong allergens. In fact, being small and refractive to degradation, they can easily pass into the blood stream, triggering immune responses.

ODORANT-BINDING PROTEINS OF INSECTS

I have already mentioned that at the same time as our discovery of the first mammalian OBP, Richard Vogt, who was working in Seattle, published a paper on the identification of a small soluble protein in the antennae of a giant moth called *Antheraea polyphemus*. This protein was produced only in the antennae of males and was able to bind the sex pheromone of this species, a long-chain acetate.

For a while we were not aware of each other's discovery. The world of insects did not have many contacts with that of mammals and we did not encounter each other at meetings or conferences until about 10 years later. In the present internet era with fast communication and easy travel such a situation would be inconceivable. However, in the past, only three decades ago, science was proceeding at a much slower and more relaxed pace. For several years we did not contact each other on the assumption that our fields of work did not have much in common.

In fact quite the opposite was the case. OBPs of insects turned out to be of great interest and were soon regarded as the insect equivalent of our mammalian OBPs. This assumption was based on several elements of similarity, all related to their functions rather than their structure. OBPs of both mammals and insects are small proteins (150–160 and 130–140 amino acids, respectively), very soluble, highly concentrated in olfactory organs, and able to bind odour molecules and pheromones.

However, when looking at the amino acid sequences, the two classes of proteins had nothing in common. Later, in 2000, when the first OBP of insects (the pheromone-binding protein, PBP, of the silk moth *Bombyx mori*), was crystallized and its three-dimensional structure solved, it was clear that its folding was completely different from that of mammalian OBPs, being constituted mainly by α-helical domains.[38] But, like the bovine and pig OBPs, the PBP of the silk moth presented a very compact structure, enclosing a binding cavity for hydrophobic ligands, in this case of the sex pheromone bombykol (Figure 24).

We have already discussed the extreme stability of mammalian OBPs and lipocalins in general and how this high refractivity to denaturation and degradation makes such proteins strong allergens. The same is true for OBPs of insects, which are even further stabilized by the presence of three disulphide bonds interwoven in a very stable network. No wonder then that insect OBPs are also endowed with allergenic activity.

At the beginning, research on these proteins in insects proceeded very slowly and was limited to species with very large antennae. In fact, the choice of the giant moth *Antheraea polyphemus* was based

Silkmoth PBP1   Locust CSP1

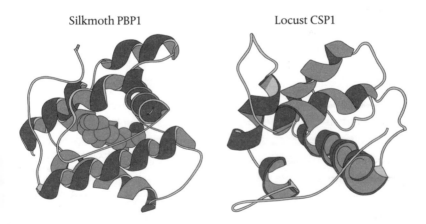

Figure 24. Three-dimensional structure of a representative insect OBP (the pheromone-binding protein of the silkmoth *Bombyx mori*) and a representative CSP (from the desert locust *Schistocerca gregaria*). In both cases the proteins are folded in very compact structures, mostly made of α-helices, but different in their shapes. In the case of OBPs, three disulphide bridges between non neighbouring cysteines further stabilize the overall scaffolding.

on the large size of its antennae. This fact and the exceptional abundance of the PBP enabled scientists to isolate more than 10 micrograms of protein from a single male moth. That might not sound very impressive, but we can do a lot of experiments with such quantities. When we compare this amount with those of other proteins in the antenna, it becomes clear that PBPs are among the most abundant proteins.

OBPs of mammals are present in the nasal mucus bathing the cilia of olfactory neurons, in a region we call *perireceptor space* and therefore they are involved in *perireceptor events*, which means anything happening to odorant molecules before they reach the cilia of olfactory neurons and meet the receptors located on their membrane.

In insects the anatomy of chemosensory organs is completely different. One of the main differences between insects and vertebrates regards the skeleton, the hard structure supporting and protecting the soft parts of the body. Vertebrates have their skeleton inside, whilst in insects it is outside, like armour. This fact has a consequence on the different strategies adopted to keep water around sensory neurons. We do not need to explain why an aqueous environment is essential for keeping the delicate terminations of neurons alive and the proteins on their surface active. Vertebrates have developed a thick mucus, a sort of jelly where very large molecules of polysaccharides bind a large number of water molecules, thus drastically reducing evaporation, in spite of the air current continuously flowing through the nasal structures. In insects, on the other hand, the ending of sensory neurons, the dendrites, are protected by a sheath of hard cuticle, which at the same time performs the important function of keeping a wet environment inside. The structure of an olfactory 'sensillum', a single sensor of the thousands located on the antenna of an insect, is drawn in Figure 25.

The hard cuticular wall of the sensillum protects the dendrites of olfactory neurons, which are bathed in a sort of thick jelly containing something like 100 mg/mL of OBPs. Producing such large quantities of proteins for tiny insects would represent a tremendous waste of energy if it was not strictly necessary for some vital functions. This

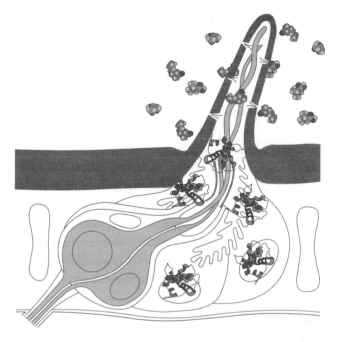

Figure 25. Schematic representation of an insect olfactory sensillum. The dendrites of olfactory neurons are encapsulated by a cuticule case filled with aqueous sensillar lymph. The main components of this lymph are OBPs synthesized and recycled by three specialized cells at the base of the sensillum. Openings along the cuticular wall let odorant molecules enter the sensillum and eventually stimulate the olfactory neurons, while water, due to its high surface tension, cannot escape.

fact, together with other pieces of evidence, strongly place OBPs among the main performers in insect olfaction but their specific roles and modes of action are still largely unknown.

Although at the beginning, research on the olfaction in insects was limited by the small size of the species, as soon as the techniques of molecular biology became widely adopted, this was not a limitation any longer and OBPs were searched for and investigated in hundreds of insect species. More recently, genome sequencing has provided a wealth of information on which to base research. In the last few years new sequencing techniques allow identification of all the genes expressed in a given organ or organism in a short time and at low cost.

Therefore, identification of new sequences is no longer the target of research, but has become the starting point; and the experimental work is aimed at understanding the function of the protein. Thus, beginning with sequences that may represent special interest, the encoded proteins are expressed in bacteria, usually in high yields, and the products used for structural studies as well as for functional investigations. Additional information on the physiological role of a specific protein can then be obtained by silencing the relative gene (thanks to more or less simple protocols) and verifying the effects at the biological, physiological, or behavioural level.

Thus, research in insect olfaction has been boosted by the introduction of all of these techniques, as well as by the practical applications in controlling the populations of agricultural pests as well as those of blood-sucking insects, like mosquitoes. As a matter of fact, research in the field of OBPs and their role in olfaction is much more active with insects than with mammals or other vertebrates. However, in spite of the large amount of structural data and functional information available, all collected with recombinant proteins, we are still to a large extent ignorant of how OBPs are involved in the detection of chemical stimuli and why such large amounts are synthesized by insects. We will come back to this question later, after having introduced the key players in the translation and interpretation of chemical signals—the olfactory receptor proteins sitting on the membranes of chemosensory neurons.

CHEMOSENSORY PROTEINS

At this point we can become better acquainted with another character involved in chemodetection. A second class of small soluble proteins has been found in the lymph of insect chemosensilla, in the same environment in which OBPs were identified. They are also present at very high concentrations and, like OBPs, are able to bind odorants and pheromones. These proteins have been named chemosensory proteins (CSPs) to indicate more generally a role in chemoreception, rather than being restricted to olfaction. In fact, at the beginning these proteins were identified in contact sensilla, suggesting a role in taste, although later they were found also in olfactory sensilla.

The structure of these proteins is composed of α-helical domains, but arranged in a three-dimensional folding quite different from that of insect OBPs (Figure 24). CSPs are also very compact and stable, a property that makes them suitable, like lipocalins and insect OBPs, for a series of technological applications.

NOT ONLY OLFACTION

The unique stability and efficiency of all three classes of proteins we have examined, mammalian OBPs, insect OBPs, and CSPs, and their ability to bind a wide variety of chemicals, matching the variety of odorants found in nature, is probably the reason why we find these proteins in many organs and tissues beside the nose or the chemosensilla.

BROADCASTING CHEMICAL SIGNALS

As soon as a purified sample of the bovine OBP was isolated, it was immediately subjected to sequence analysis in order to get some indication of the class of proteins to which OBP could belong. Even before the era of the genomes, there was a large database with protein sequences and it was likely that we would find some similarity with already described proteins. In fact, it took only a short segment of sequence of about 25 amino acids to reveal a marked similarity with a class of small soluble proteins purified from the urine of mice and rats.

OBPs carry pheromones in the urine of rodents

These proteins, named MUPs (major urinary proteins) had been described in 1965, but, two decades later, their function still represented an unsolved mystery. It was a disturbing fact that MUPs were excreted in the urine of mice at a rate of up to 10 mg per day, representing about 10 per cent of the nitrogen balance of the animal. Such a large waste of energy had to be justified by a very important function.

Curiously, the great similarity between MUPs and OBPs, instead of throwing light on the function of OBPs, provided a reasonable explanation for the presence of MUPs in urine and a plausible solution to the 20-year old riddle. It was easy to reason that if OBPs bind odorant and pheromones, MUPs would bind such ligands too. This was later experimentally demonstrated by Andrea Cavaggioni, who also found, much more interestingly, that when MUPs were purified from the urine of mice they were loaded with chemicals that had been previously identified as the mouse pheromones.

So the task of these proteins was to bind pheromones and keep them in the aqueous medium of urine. Nearly all animal species mark their presence with pheromones and several of them use urine as a carrier. If a pheromone, because of its hydrophobic nature, also linked to its volatility, does not dissolve well in water, then a protein helps, providing a hydrophobic environment in its binding cavity. But there is more: being inside the protein, the pheromone molecule is protected from degradation by environmental agents and its volatility is decreased. In other words, by encapsulating the love message in a protein, the mouse ensures a longer life to its advertisements for potential partners.

OBPs as pheromones

Is that enough to justify the use of such an expensive container for a love gift? Probably not. Carla Mucignat, who was working at that time with Andrea Cavaggioni, wondered whether the protein itself could exhibit some pheromonal activity. It should be noted that MUPs are strictly male-specific and are under hormonal control: they are only found in adult mature males. Therefore their function should be searched for within the reproductive area.

Carla injected the purified protein without any ligand into the nasal cavity of young female mice and observed that in the individuals treated with the MUP the reproductive organs matured faster than in the control mice. This experiment indicated that not only the bound ligand, but the carrier protein itself acted as a pheromone. The MUP is not just a box for a gift, it really *is* the precious gift.[39] This interpretation also explains the observed behaviour. When a urine mark is left by a male mouse, females find their way guided by the scent gradually released into the air. Then, once the urine spot is found, the female starts licking and sending the protein into the vomeronasal organ, a special cavity beneath the nose and connected to the mouth, dedicated to the perception of pheromones.

As the work proceeded and more sequences were obtained both of OBPs and MUPs, it became clear that in fact both these groups of proteins belong to the same class and some of them are found in the nose as well as in the urine. This is not so surprising if we regard pheromonal communication as a broadcasting station (the urine) and a receiving apparatus (the nose), using the same wavelength (the binding protein) for sending and detecting the same message.

We are well aware now that such a double system is widespread in nature and often OBPs are endowed with this dual function. It is not always the urine, however, which is used as a medium to broadcast pheromones. Other biological fluids are engaged in activity such as the vaginal secretion in hamsters, the sperm in rabbits, the sweat in horses, and the saliva in pigs. Each animal species adopts a different secretion as a means of transporting pheromones and in all these secretions where there is a pheromone we also find an OBP.

OBPs and pheromones in boar's saliva

I have already mentioned more than once the pig pheromone androstenone, which is released in the saliva of the boar and poses serious problems to the quality of meat. Androstenone is a highly hydrophobic compound and badly needs a protein to stay in the watery environment of saliva. There are actually two OBPs produced in the sub-maxillary glands of the boar and they contain in their binding cavities the two components of the boar pheromone, androstenone and its corresponding alcohol androstenol. Both OBPs are also present in the nose, but there they are void of ligands. As observed for the MUPs, these two OBPs are male-specific in the glands, but are expressed in the nose of both sexes.

Insect OBPs and CSPs as pheromone carriers

What about insects? Do we observe a similar phenomenon in which the same proteins release and detect pheromones? This is indeed the case, and is well documented in several species. Insects possess pheromone glands which are clearly identifiable and easy to dissect. In many species glands producing sex pheromones are located at the far tip of the abdomen and are usually hidden. During the process of *calling*, a female extrudes a translucent little ball impregnated with pheromone, which is soon released into the environment. In several species of insects, OBPs as well as CSPs have been detected in these glands. Not merely one or two, but sometimes a dozen or more. Clearly the main function of these proteins in the glands is to make the pheromone soluble and perhaps regulate its gradual release in the environment, more or less like the MUPs in the urine of mice. But why do we need so many? Are they also acting as pheromones like the MUPs? Currently, we do not know the answer to this question.

Besides such glands producing sex pheromones and being present in virtually all insect species, there are other pheromone glands in different parts of the body, particularly when we look at social insects. We have already observed that in these species the chemical language is richer because individuals need to communicate with each other sending precise information about their identity, foraging sites, and presence of danger, as well as giving orders and detailed instructions for specific tasks, such as building a nest, attacking another nest, caring for larvae, and many other operations.

In the honey bee, for instance, mandibular glands are highly developed and produce several pheromones. The same glands also contain a number of OBPs and CSPs. The synthesis of these proteins is regulated according to caste and age. In particular, as the workers are assigned different tasks during their life, first tending the larvae, then looking after the hive, fighting intruders, and finally collecting food, their expression profile of OBPs and CSPs is also modified accordingly.

Obviously there is much more to these proteins than just solubilizing pheromones. One hypothesis, which still needs to be verified, is that binding proteins might regulate the composition of the pheromone blend. In other words, instead of adjusting the synthesis of the different components, it is easier to regulate the expression of their binding proteins. In this way, the composition of the pheromone mixture will be determined by the relative amounts of the binding proteins. The synthesis of a pheromone component requires the action of several enzymes, whose synthesis should be activated when needed. By contrast, the synthesis of a carrier protein is the direct effect of the activation of a single gene. Moreover, OBPs and CSPs are small proteins, which spontaneously fold into the correct structure and do not need any other factor, as enzymes often do, to bind their ligands.

OBPs and CSPs have also been identified in the reproductive organs of insects. In the cotton bollworm, an OBP which is highly expressed in the antennae is also present in the sperm. This protein is transferred to the female during mating and ends up on the surface of eggs—only fertilized eggs of course. As in other cases, this OBP carries endogenous ligands, probably pheromones whose function still waits to be clarified.

A similar case is that of the yellow fever mosquito *Aedes aegypti*, where an OBP is transferred to the female in the sperm, while in the oriental locust, *Locusta migratoria*, as many as 17 CSPs have been identified in female reproductive organs, but only one in the sperm. We are just beginning to understand how complex chemical communication between sexes can be even in relatively primitive insects. We assume that each of these OBPs and CSPs would be a carrier of specific pheromones to modulate courtship and all the phases of reproduction. But there might be more than chemical communication going on.

The role of OBP and CSP carriers for lipophilic ligands extends beyond pheromones and odorant molecules. I have described the case of retinol-binding protein, a lipocalin ferrying the highly hydrophobic retinol across the blood from the liver to the retina. A similar situation occurs in insects: OBPs and CSPs have been found in the eyes of cotton bollworm and of the honey bee likely as carriers of hydroxy-retinal, the oxidation product of hydroxy-retinol used in vision. The phenomenon is most likely present in all insects. The same or similar proteins are recruited to transport pheromones and visual pigments. Despite their different functions, these ligands share a physico-chemical characteristic; they are highly hydrophobic and need to be encapsulated into protein shells so as to travel across aqueous physiological media.

We have seen that the urinary proteins of mice are physiologically active on the maturation of young females. At least in one case it has been demonstrated that a CSP is required for the maturation of the embryo. The CSP3 of honey bees is only found in ovaries and eggs and nowhere else in the body, a rather peculiar fact, as CSPs are often found in several organs. When the gene encoding this protein was silenced, thus suppressing the synthesis of CSP3, the embryos did not develop correctly and were not able to emerge from the eggs.

Another CSP is certainly involved in the regeneration of legs in the cockroach. Curiously, this protein and its actions were reported a few years before CSPs were identified in sensory organs and studied as odorant and pheromone carriers. If you cut a leg from a cockroach during its larval stage, the insect is able to regenerate a whole limb. During such a process, the synthesis of a protein, named at that time as p10, but with a sequence very similar to CSPs, increases dramatically, to return to normal levels when regeneration is complete.

Besides such important and vital tasks of OBPs and CSPs, these proteins are sometimes utilized for apparently humble requirements. The proboscis of the cotton bollworm and related species contains

exceptionally large amounts of CSPs. There is good evidence supporting a role for this protein as a wetting agent to ease the flowing of
liquids across the proboscis and reducing the effort needed for sucking. All proteins act as sorts of detergents, lowering the surface tension
of water and wetting the surfaces in contact with solutions. It is
common practice to wet the glass of your underwater mask with
saliva to prevent the formation of water droplets on the inner surface.

PROTEINS FOR MANY TASKS

The number of OBPs and CSPs involved in tasks other than chemical
communication is continuously increasing, as researchers widen their
field of investigation and eliminate restrictive assumptions and
hypotheses which often in the past represented obstacles in the
progress of research. In fact, the idea that a particular tool, like a
protein, should only be used for a specific task, within a specific
organ and a specific physiological function is by no means supported
in biology, nor could it be taken as a reasonable hypothesis. On the
contrary, once a particularly efficient tool or mechanism has been
obtained through evolution, it is protected from further changes.

Probably the best example is rhodopsin, the protein at the basis of
vision, which is used to detect light from primitive algae right up to
humans with only minor changes. At the same time, such efficient
tools are adapted for other functions with only limited modifications.
Thus a superfamily of proteins is generated. That many proteins are
organized in superfamilies is the best evidence of the fact that when a
successful structure has been obtained, it is utilized for many often
unrelated tasks. In this respect, the superfamily of lipocalins is probably the best example of this phenomenon.

We have seen that the superfamily, to which vertebrate OBPs
belong, also includes other lipid-binding proteins with different functions, such as retinol-binding protein and fatty acid-binding protein.
Even some enzymes and membrane bound proteins have been classified within this family, as well as β-lactoglobulin, an abundant

protein in milk, whose function is still unknown. What all these proteins have in common is some sequence features, but most important a common compact folding reproduced with high similarity even in members that are distantly related when you only look at their amino acid sequences.

A scenario similar to that of vertebrate OBPs and lipocalins is also gradually being revealed in the world of insects. Both OBPs and CSPs appear more and more as merely the tip of the iceberg, representative of larger families of proteins endowed with different functions both in chemical communication and beyond.[40]

9

RECEPTORS AND BEYOND

From Odorants to Emotions

A LONG AWAITED DISCOVERY

They certainly existed, we were all convinced of that and we also knew what they would look like, yet they still managed to escape all attempts to find them.

The identification of olfactory receptors was the result of a series of events which together contributed to make their search practical and feasible. Certainly the discovery of OBPs showed for the first time that a biochemical approach could be applied to study olfaction and at the same time provided the right tools and indicated paths to follow. On the other hand, molecular biology, a discipline rapidly growing, offered new efficient tools to study the genes, the sequences of DNA encoding proteins, thus showing a short-cut to the identification of receptors. In this process, perhaps the most important contribution was provided by the introduction of new (at that time) techniques for amplifying DNA sequences and producing billions of copies in a simple, rapid way. This proved of utmost importance in the study of genes present in a small number of copies, as those encoding olfactory receptors.

In fact the tiny amounts of these receptors present on the olfactory epithelium was the main reason for the failures experienced by several groups of researchers during the preceding decade, who, stimulated by the success obtained with OBPs, had been searching for the holy grail of olfactory receptors. Employing biochemical tools, only poor and dubious results had been obtained from time to time, which however could not be reproduced by other research groups. A scientific discovery needs to be confirmed and other research groups should be able to reproduce the same results before it gets credit and is accepted by the scientific community. This was the case for OBPs, which had aroused immediate interest as soon as our results were reproduced by American colleagues.

We have observed how the tiny amounts expected for membrane proteins already represented a major difficulty for their identification. But even more problematic was dealing with a large number of receptors, several hundred expressed in each species, according to our present genome information, all differing from one other, but so similar in chemical properties as to make their separation virtually impossible.

In fact, the extreme complexity of the olfactory code had been so far overlooked by most scientists. To be simple and elegant, like the colour vision code, the olfactory system was assumed to be based on a small number of elementary sensations, a view that would soon be proved totally wrong.

Contrary to such common and widespread beliefs, Richard Axel and Linda Buck were convinced that olfactory receptors constitute a very large multigenic family—a brilliant intuition that proved to be the winning card. In such a scenario, the idea of searching for the receptor proteins, using biochemical methods was out of question, and the two scientists directed their attention to the corresponding genes.

Towards the end of the eighties, when Buck and Axel started their research, molecular biology was an established discipline and the technique of PCR in particular was gaining popularity for its unprecedented power in studying genes expressed at very low levels, as those of olfactory receptors were expected to be.

Amplifying a gene through PCR is very easy if we already know at least some parts of its nucleotide sequence. But for olfactory receptors it was a different story. You had to search among hundreds of thousands of sequences without really knowing what you were looking for. In fact, there was only a very weak thread to follow: it was known that a special membrane protein, called a G-protein, was a stepping stone on the way from olfactory stimuli to perception. It was also known that G-proteins were coupled to a special family of receptors embedded in the the cell membrane, which they cross seven times, winding alternately in and out of the cell. It is for this characteristic that they are referred to as the 7-TM (seven transmembrane) family. Rhodopsin, the protein that forms the basis of vision, and β-adrenergic receptors (mediating detection of neurotrasmitters) are members of this family and they had already been well studied at that time. Therefore, Buck and Axel focused their attention to these receptors, whose sequences were chosen as the templates to design short fragments of DNA (primers) to be used in PCR experiments in the hope of hooking some of the olfactory receptors.

It looks simple and straightforward, but in fact searching for olfactory receptors is like hunting for the proverbial needle in the haystack. Several laboratories at that time were following the same track, performing hundreds of PCR experiments, cloning and discarding genes all the time, like a gold rush, in the hope of finding the dreamed-of nugget, that single sequence which would show the way to all the olfactory receptor genes.

In such cases what you need most is patience and perseverance, the capacity for enduring failure day after day without giving up, and determination to find the ultimate object of your search. Of course you should firmly believe that what you are searching for does exist and that eventually you will find it. Linda Buck had all these qualities. Perhaps she was better than others at designing primers and setting up experiments, perhaps she was luckier, but certainly what determined her success at the end of three years of failed experiments was tenacity.

Linda Buck and Richard Axel published their first results at the beginning of 1991 in the prestigious journal *Cell*.[41] The publication of their paper came like a bombshell. It was like breaking down a wall and gaining access to the new and unexplored world that was concealed behind it. Thirteen years later the importance of their discovery was recognized with the Nobel Prize for Medicine or Physiology being awarded to the two scientists.[42]

It was obvious to everybody working in olfaction that those fragmented sequences reported in the *Cell* paper held the code to understanding the language of odours and contained the key which opened many doors. Some laboratories which had already been searching for olfactory receptors took immediate advantage of the new information to jump ahead. Other groups changed their approaches to olfaction and built new labs of molecular biology to focus on receptors. What only a decade earlier was regarded as a risky and insecure area, where it would have been advisable not to venture, appeared now as the Holy Land, a virgin field capable of rewarding researchers with many interesting fruits.

For the first time scientists had direct access to those receptors responsible for reading chemical information encoded in the structures of odorants and translating them into electrical signals which gave rise in the brain to emotions, verbal expressions, and behavioural reactions. The first and most important brick had been laid, but a lot of research was still waiting to be done and all the mechanisms leading to perceived sensations in the brain with the intricate neuronal network through which olfactory messages travel, interfere with each other, and interact with other areas of the brain, had yet to be understood and clarified.

Olfactory receptors represent the largest multigene family

Soon after the discovery of olfactory receptors, the first question which stimulated the curiosity of scientists was their number. It was already clear that the number of genes encoding these proteins was far higher than any prediction that had been made on the basis of psychophysics studies. On the basis of the first data, it was possible to place their number in rats at around 1000, making olfaction closer to hearing than to colour vision in terms of the complexity of the code.

Later, when genome sequencing provided access to all gene sequences the number of olfactory receptors in the rat appeared close to 1500, of which about 20 per cent are not functioning. These are called pseudogenes because they contain some errors in their sequence preventing them from being expressed. Similar situations were discovered in the mouse and in the dog, with about 1300 and 1100 genes, respectively. Also in these species around 20 per cent of the genes, still remaining in the genome, are not expressed. When we look at humans, the situation is different. Although the total number of genes is more than 900, fewer than 350 are still intact and potentially active. This fact clearly indicates that we are progressively losing our sense of smell.

How and why is this happening? During evolution, random mutations in the genes occur continuously owing to errors. The imperfection of the gene replicating machine, rather that representing a problem, proved to be beneficial and of extreme importance for evolution. In fact, it is thanks to casual errors that evolution occurs at all. A large proportion of these errors have no consequence for the protein encoded by a particular gene, as the genetic code is redundant and in many cases the substitution of a single base may produce a triplet still encoding the same amino acid. When, on the other hand, the amino acid is changed, the mutation can be beneficial, neutral, or detrimental for the life and health of the individual. Natural selection will then choose which condition is more favourable and in the end this will be the winner. In some cases, the substitution of a single base can lead to a stop codon, that is, a signal telling the system to interrupt the synthesis of the protein. Or else it could lead to other problems, all having as a consequence the failed expression of the protein. It is clear that if that specific protein is either essential or very important for life, the individual may die before being able to pass on his or her genetic heritage to offspring. But, if the absence of that protein does not affect the life or the health of an individual, this faulty gene is transmitted to further generations, thus spreading the mutation around.

Humans do not rely too much on the sense of smell. Even a completely anosmic person can lead a normal life and would not suffer for this anomaly when finding a partner or selecting food. Moreover, given the large number of olfactory receptors, the presence of a handful of non-functioning receptors passes completely unnoticed during the life of an individual. This explains the high incidence of pseudogenes in the human population, as compared with other mammals. When we look at primates, this trend is confirmed with around 30 per cent of pseudogenes in lower monkeys to more than 50 per cent in apes. In the human population, we can reasonably assume, and to some extent psychophysical studies have supported this view, that each of us is lacking several olfactory receptors and we cannot speak of a *normal* subject in olfaction, as we do when dealing with colour vision.

Olfactory receptors across vertebrates

When searching for olfactory receptors in new species, one fortunate characteristic of these sequences is the absence of *introns*. Usually a gene contains coding regions (*exons*) interspaced by other segments (*introns*) which are cut off when assembling the RNA from DNA. Therefore, at the RNA level the gene, which will be translated into a protein, derives from stitching together different fragments which in the genome may be located quite distantly.

The olfactory genes of vertebrates, however, do not undergo any editing. This means that we can collect our sequences directly at the genome level, making the search much easier. Besides, olfactory receptors are relatively well conserved across vertebrates, so that for instance, based on the rat sequences, olfactory receptors have been identified with relative ease in other species of vertebrates.

For example, the zebrafish contains in its genome 133 sequences encoding olfactory receptors, of which less than 100 are functional, while in the tropical frog *Xenopus tropicalis* the number of the genes is close to 900 with about 400 functioning. There is a curious observation about this frog, that is worth reporting.

Amphibians lead a double life, aerial and aquatic, and *Xenopus* is no exception. To cope with this double personality and be sure to detect the different environments efficiently, this species is equipped with two noses. Strange as it may sound, they have two nasal cavities which they can open and close according to the environment in which they are located. There are two populations of olfactory receptors present in the two noses. Those of the aquatic nose are more similar in amino acid sequences to those of fish, while the others are more similar to olfactory receptors of mammals.

Another unusual situation is found in the chicken, in which of a high number of genes (around 550) only 80 or fewer are active, as in humans. In fact, the role of olfaction in birds has long been a matter of debate and the existence of pheromones is still questioned. Birds are known for their excellent vision—think of the expression, having an eagle eye. Certainly many aspects of bird life are determined by vision, as witnessed by the bright colours of the plumage in many species, usually arising from sexual selection, and shown off by males to court females. But olfaction is far from being absent in birds. The best known example is provided by homing pigeons, which can travel long distances and find their nest on the basis of an olfactory map they have assembled in their brain. Other migratory birds have also been shown to use olfaction for their orientation and in some species putative pheromones have been identified.

What happens when air-breathing mammals return to water, as in the case of whales and dolphins? The aerial nose is not useful any longer and apparently they have not developed and cannot develop at this point in evolution an aquatic nose. As a matter of fact, cetaceans are anosmic. They are one step ahead of humans having already lost all their olfactory receptors. A few olfactory receptor genes have been found in whales, but they appear to be all pseudogenes.

From men to worms

Worm is the nickname given by scientists to *Caenorhabditis elegans*, a very small nematode, barely visible to the naked eye, which has, for over half a century, been one of the most useful and interesting models for the study of genetics and development. It was also the first multicellular organism to have its genome sequenced in 1997. More recently this organism was adopted by neurobiologists as a model for examining the organization of the nervous system, thanks to its extreme simplicity. In fact it is made of only 302 neurons, yet still a large part of the 959 cells that constitute the entire body of this worm.

Despite such simplicity, the olfactory gene repertoire of *C. elegans* accounts for more than 500 functional genes.[43] But all these chemosensory receptors are packed into 32 neurons. This means that each neuron houses several receptors and therefore has the ability to detect a certain number of chemicals. However, it cannot distinguish as different all the smells reaching the same neuron, that have to be grouped in the same category. Given its limited resources, the worm has developed an efficient system combining the capacity to detect a large number of different chemical structures with its very simple anatomy.

What is important for the individual is how to react to a stimulus, rather than identifying the nature of the stimulus. Therefore, to give just one example, potentially toxic compounds fall into different chemical categories, needing a large repertoire of receptors to be accurately detected, but what is important is that all of them give rise to the same signal of alarm so they can be housed in the same chemosensory neuron, eventually triggering an avoidance response. In a similar way, another sensory neuron can respond to different types of food; yet the system is not able to discriminate between different flavours, but as long as the food is edible, the message instructs the worm to move forward. This is, in a way, like a traffic light sending simple signals and commands, without the need to explain them.

OLFACTORY RECEPTORS ARE TRANSMEMBRANE PROTEINS

It is time now to introduce these central characters of olfaction and take a look at their structure. Olfactory receptors are proteins consisting of little more than 300 amino acids, containing seven hydrophobic segments which traverse the cellular membrane, just like rhodopsin, β-adrenergic, and others receptors. The terminal amino group of the chain (considered as the starting point) is located outside in the extracellular space, while the C-terminus (the ending) is found inside the cell.

The cell membrane, for those who are not familiar with its structure, is a double layer of phospholipids. These are strange compounds containing a hydrophobic tail, made of two long-chain fatty acids and a hydrophilic head constituted by a molecule of phosphoric acid linked to a small organic molecule such as choline or others. These three acids are held together by a molecule of glycerol. They are similar in a way to triglycerides, the molecules constituting most of our edible fats, such as oil or butter, in which a molecule of glycerol is connected to three fatty acid chains.

But, owing to the presence in the same molecule of a phosphoric acid group and a long fatty acid chain, phospholipids exhibit a strange behaviour, being at the same time hydrophilic (because of the phosphoric *head*) and hydrophobic (because of the fatty acid *tail*). In an aqueous environment, these molecules can easily self-assemble into double layers, where the heads interact with water, while the tails interact with each other (Figure 27). Fragments of such films can fold into spheres, thus delimiting a closed area and an outside environment, both containing water, but not able to communicate with one another because of the lipid barrier. In this way the first cells were generated, which was an important step towards the development of life. In fact, the membrane gives an entity to the cell, a unit defined by a physical barrier, that is capable of replication.

But the cellular membrane is much richer and more complex than this. Proteins, like our olfactory receptors, sit across the walls of this tiny city, checking on every visitor approaching the wall and sending appropriate messages to the inside. Gates can be opened and closed following chemical instructions and admit ions or other molecules.

Structure of olfactory receptors

Olfactory receptors interact with this lipid barrier by spanning it seven times, winding in and out of the cell (Figure 27). Therefore, they contain hydrophobic amino acids in the regions that have to interact with the membrane. Mentioning interactions is a bit like talking of solubility. Polar compounds dissolve better in water, fats dissolve better in oil or organic solvents. Therefore, we can easily recognize the segments along a protein sequence which are rich in hydrophobic amino acids as those most likely to be crossing the cellular membrane. And we can count the number of transmembrane regions and assign a particular protein to its class. In this way, the simple information from the amino acid sequence can already reveal some characteristics of an unknown protein. This is quite important in our era of genomes, having reached a stage where sequence information can be easily obtained long before we can hypothesize a physiological role for a protein.

Let's now go back to the structure of olfactory receptors to get more insight into how they could recognize different odorant molecules. All receptors belonging to the 7-TM family share the same type of compact structure. So far only the three-dimensional shape of rhodopsin and a couple of other receptors has been experimentally solved, therefore all models and hypotheses are mostly based on what we know about rhodopsin. It basically consists of seven segments of α-helices, the regions crossing the membrane, packed together like a bundle of pencils, connected through loops of non-ordered structures alternately bathed in the extracellular or intracellular fluid. This tight assemblage of the seven helices, however, leaves a channel inside, where the odorant molecules are thought to be captured and recognized.

At least this is where the permanent ligand of rhodopsin, retinal, is entrapped. Retinal is an aldehyde of 20 carbon atoms of rather complex architecture, reproducing half of the structure of β-carotene, a pigment widely present in plants. Owing to the presence of a large number of alternating double bonds, it can absorb light in the visible region. When this happens, the chain of retinal undergoes a major twist around one of the double bonds. As retinal is strongly hooked to the molecule of rhodopsin through a covalent bond, such a twist induces a conformational change in the protein, which then communicates a like change inside the cell by interacting with a G-protein. Figure 26 shows the structure of rhodopsin and the two forms of retinal.

Although rhodopsin and olfactory receptors appear very different in their functions, the first detecting light, the second volatile

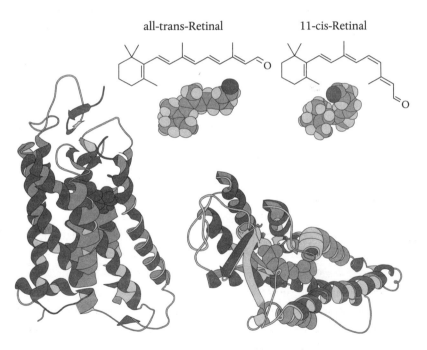

Figure 26. Two views of bovine rhodopsin bound to a molecule of retinal. Retinal undergoes isomerization from the all-trans form to the 11-cis isomers when hit by a photon. This produces a conformational change in the bound rhodopsin, which eventually generates an electrical signal in the cell.

molecules, they nevertheless share more or less the same mechanism. In fact, odorant receptors *sense* the presence of a foreign molecule interacting with its core and changing its conformation, while rhodopsin detects the twist induced by light in the molecule of retinal and changes its conformation. Observed from this perspective, rhodopsin is also a chemical sensor, if we consider the two forms of retinal as two different molecules, as indeed they are.

FROM MOLECULAR INTERACTIONS TO ELECTRIC SIGNALS

When odorant molecules reach the surface of olfactory neurons, they are promptly blocked by olfactory receptors, sitting across the membrane like sentinels to check visitors and reporting on their presence to the inside of the cell. Recognition by receptors occurs on the basis of shape, size, and other chemical characteristics of the smelling molecules. Now the chemical information encoded by these specific interactions between odorants and receptors has to be translated into an electric signal, which can be more easily measured, amplified, and processed just like electric currents in the circuits of a computer or any electronic instrument. This translation is accomplished by a series of enzymatic reactions triggered by a change of conformation that the receptor undergoes when accepting a small organic molecule, like an odorant, inside its structure.

Olfactory receptors send messages inside the cell

The first biochemical element to detect that an odorant molecule has been captured by an olfactory receptor is a G-protein, a complex enzyme made of three subunits, which is in physical contact with the receptor. The G-protein, when stimulated by the conformational change of the receptor, initiates an enzymatic cascade, a series of chemical reactions leading to the production of large quantities of cyclic AMP. This soluble molecule travels across the body of the cell (in our case an olfactory neuron) and binds to ion channels, opening

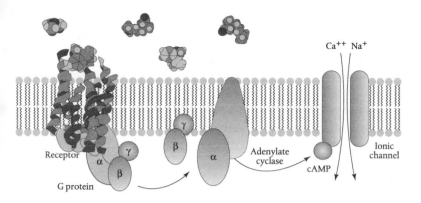

Figure 27. Main steps in olfactory transduction. The interaction of an odorant molecule with the olfactory receptor, induces a G protein to dissociate and activate the enzyme adenylate cyclase. The product of this reaction, cyclic AMP (cAMP) triggers the opening of an ion channel, leading to depolarization of the neuron and generation of an electric signal.

them just as a key opens a door. Figure 27 illustrates the main events of olfactory transduction.

These channels are proteins of very complex structures, acting like holes in the membrane through which specific ions can travel. Thus, the opening of these channels produces a flow of ions from outside to inside or vice-versa, resulting in a depolarization of the cell. In practice, the electric potential of the cell, due to an imbalance of ions between the interior and the exterior, is rapidly reduced because of the flow of ions. Looking at the process as a whole, the chemical interaction of an odorant with its specific receptor has generated an electrical impulse. The translation of a chemical message into electric signals which can be further amplified, processed, and compared, is the key step connecting the external environment to the brain.[44]

SPECIAL TOOLS FOR SMELLING PHEROMONES

The vomeronasal organ, which we introduced earlier, is a small cavity, usually a blind-ended passage, present in most vertebrates and dedicated to detection of species-specific pheromones. It is practically a second nose or a third major chemoreception organ. Like the nose and the tongue, this area is equipped with receptors, still G-coupled 7-TM proteins. Actually there are two classes of such receptors in the vomeronasal organ, classified as V1R and V2R. Receptors of the first class are more similar in size and structure to olfactory and gustatory receptors, although their amino acid sequences are very different.

The receptors of type V2R, instead, present, in addition to the region containing the seven transmembrane helices, another domain as large as the core of the protein, extending into the extracellular space. This region of the protein has been suggested to be a potential binding site for pheromones of protein nature. In fact, proteins with pheromonal properties, well known in yeasts and in reptiles, seem to also be utilized by mice.[45] We observed in Chapter 6 that MUPs, the urinary proteins of mice, can trigger physiological changes in young females leading to early maturation. More recently, another member of the MUP family, named 'darcin' (after Mr Darcy, Jane Austen's hero in *Pride and Prejudice*), has been reported to be endowed with pheromonal activity.[46]

What about humans? Do we have vomeronasal receptors? The question is important, because it is related to the possibility of pheromonal communication in humans, and will be discussed further in Chapter 10.

THE TASTE OF FOODS

We have already noted that unlike the complexity of the olfactory language, taste is based on a very simple code, made of just five letters. All taste sensations can be classified as sweet, bitter, salty, acid, or umami. Such simplicity has been also observed at the level of receptor proteins dedicated to detecting such sensations. We find again 7-TM receptors, similar to the olfactory ones and also coupled to G-proteins. The chain of events leading to opening ion channels and thus generating an electric potential is also similar to that described for olfaction.

But the system is much simpler with a single receptor for sweet compounds, one for umami, and a handful for bitter tasting substances. The perception of salt and acid then is even more basic, as in such cases ion channels are directly activated by salts and acids.

Bitter is a warning signal

A bitter taste is a warning signal related to potentially toxic compounds. Avoidance behaviour regarding bitter tasting food has certainly evolved because those who were unable to perceive the taste of toxic plants or experienced them as pleasant ended their life before they could pass their genes to the next generation. There are thousands of naturally occurring alkaloids and other bitter compounds.

Gustatory receptors vary in number across species from as few as three in chickens to around 50 in amphibians. Humans exhibit at least 25 such receptors named T2R (taste receptors of class 2) followed by a numeral. They are housed in the taste buds of circumvallate papillae, located at the back of the tongue following a strategy similar to that observed for chemoreception in *C. elegans*. Rather than having a single type of receptor protein per sensory cell, as in olfaction, we find several bitter receptors in the same taste bud. The information becomes blurred and non-specific, but the message is clear: keep away from such chemicals. The advantage is a much simpler system.

Then why, one could ask, do many of us appreciate the bitter taste of chocolate, coffee, or some vegetables and digestive drinks? As we have already observed, we have learned that some bitter tasting foods are safe to eat and have also learned, as a result of cultural education, to appreciate them. In fact, children do not like anything tasting bitter, while they are attracted since birth to sweet food. How, then, can we distinguish between different shades of bitter taste? Although several receptors are housed in the same sensory bud, not all of them are packed together, so that some discrimination is still feasible. However, what lets us discriminate between bitter chocolate and unsweetened coffee is more likely to be the presence of olfactory notes accompanying the basic taste sensation, which enrich the flavours and make each one of them unique.

Sweet and umami are indicators of good food

A sweet taste and the perception of umami, the typical characteristic of glutamate dominating in meat broth, are detected through an even simpler system. Altogether we are equipped with only three receptors of type 1, defined as T1R1, T1R2, and T1R3. These three elements form heterodimers, representing the real active receptors. The combination of T1R1 and T1R3 detects umami, while the dimer T1R2 and T1R3 detects sweet-tasting chemicals.

The unexpected fact is that a single taste receptor can deal with the large variety of sweet compounds. Sugars, like glucose or sucrose are structurally very different from saccharine, aspartame, or cyclamate, artificial sweeteners 50–300 times sweeter than sucrose. Proteins also can be extremely sweet, like thaumatin and monellin, which can be thousands of times sweeter than sucrose. All these structures which differ in size, polarity, and chemical groups certainly need a range of differentiated receptors. This was the accepted idea which molecular biology and genome sequencing disproved beyond any doubt, by demonstrating that a single receptor existed for all such chemicals. Molecular modelling has later shown how such a variety of structures can all interact in efficient ways with the same sweet receptor.

Salts and acids follow direct channels

The other two taste modalities, salty and acid, are not perceived through any receptor of the types examined so far. They are simply ion channels to measure the concentrations of both salts and acids. When we eat a salty dish, the concentration of sodium ions outside the cells on our tongue becomes higher than inside. Specific ion channels open to let sodium ions inside, thus balancing the concentrations on the two sides of the membrane. This creates a change in the electric potential of the cell, acting as a signal.

Similarly we detect the presence of an excess of hydrogen ions through a signal that we perceive as acid: it is a sort of pH meter we have on our tongue. Therefore, our sense of taste is not able to discriminate different acids, such as acetic acid in vinegar or citric acid in lemon, but only measures the strength of acidity. Of course we can recognize lemon from vinegar and we do appreciate different types of vinegar, but once again all this has to do with olfaction. Without the accompanying volatile chemicals which stimulate our sense of smell all acidic compounds would elicit the same sensation.

For those who want to learn more about the molecular aspects of taste perceptions, several excellent studies have been published.[47]

NOT ONLY IN THE NOSE

One of the important guidelines that helped Richard Axel and Linda Buck in their successful search for olfactory receptors was the assumption that such genes would only be expressed in the olfactory tissue or at least in chemosensory organs. But, soon after their discovery, a paper reported the occurrence of some olfactory receptors in the sperm cells.[48] This disturbing finding was not easy to accept and the first hypothesis was that the experiments had not been well performed. However, the data were clear and sound and were confirmed more than once by subsequent studies.

Once this fact was accepted, it was easy to guess what olfactory receptors do in sperm cells. We have already observed how olfaction is only one aspect of chemoreception. Cells communicate with each other and with the environment and the very formation of an organism is based on such a network of messages exchanged all the time between the elemental components of the organism. Cells get together to form an organism and need to talk to each other in order to assign and assume different roles and functions. In this way, cells start differentiating in order to build a complex organism. In a similar way, ants exchange chemical messages to assign different tasks to each other and perform specific roles within the large community of the nest, functioning as a sort of a superorganism, as we discussed in Chapter 5.

Sperm cells are guided by smell

Sperm cells need to find their way to the egg. It is a long, complex, difficult quest, and highly competitive. We know that only one cell out of the millions which embark on their journey will be successful. An efficient compass (or, better, navigator) is essential to find the shortest way. It is a chemical compass and sperm cells swim towards their goal guided by their *nose*, just like bacteria swimming towards a food gradient in a process called *chemotaxis*.

So, why do they use olfactory receptors instead of other receptors? Well, we could ask why shouldn't they use olfactory receptors, which work so efficiently and have their genes already present in the DNA? We have seen, in other instances, how nature tends to use the same tools for different functions, once they prove efficient and versatile enough.

The next question is: what do sperm cells smell with these receptors? Perhaps the term 'smell' is not really appropriate, but in any case they are sensing the presence of certain molecules, signposts showing the correct way to the egg. It might appear strange and surprising that in the era of genomes, computers, and nanotechnologies we still don't know what the egg smells like to the sperm cells. There are certainly molecules released by the egg or by its closed environment to attract sperm cells, otherwise they would not be swimming so fast and heading in the right direction.

Hanns Hatt, one of the pioneers of olfactory research, and his collaborator Marc Spehr, at the University of Bochum in Germany tried approaching this question in an indirect way. They focused their attention on a specific human olfactory receptor which they managed to express in some cell lines and identified a series of chemicals able to activate this receptor. The best ligand, by a strange coincidence, proved to be bourgeonal, a synthetic perfume ingredient endowed with a lily-of-the-valley scent and present in many fragrance

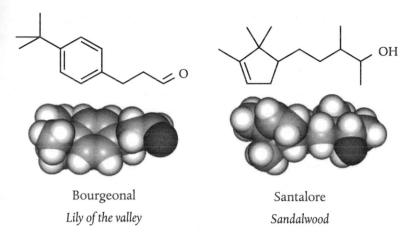

Bourgeonal

Lily of the valley

Santalore

Sandalwood

Figure 28. Ligands of olfactory receptors not involved in chemoreception. Bourgeonal is a good ligand for an olfactory receptor expressed in sperm cells and probably involved in chemotaxis towards the egg. Santalore has been shown to bind to another human olfactory receptor involved in skin cancer.

formulations (Figure 28). They also demonstrated that this odorant can stimulate the receptor present on live cells by monitoring the entrance of calcium ions through ion channels opened by the responses of receptors to odorants.[49] Later, other studies demonstrated that sperm cells swim towards this odour. A similar story was also reported in mice. This is as close as we are to identifying the natural smell compound which drives sperm cells frenetically towards the egg, but its chemical nature still remains unknown.

Olfactory receptors and cancer

But olfactory receptors in sperm cells still had another surprise in store for us. There is good evidence that they can help to control prostate cancer growth. Another olfactory receptor, different from the one discussed above, is found both in the prostate and in the nose. This receptor has been reported as a prostate tumor specific biomarker, as its expression increases in such cancerous tissues. The same Hanns Hatt group, which studied the sperm receptor, decided to look deeper into the mechanism of action of this prostate member and found that it binds some steroids as well as some terpenoids. In particular, β-ionone, a violet smelling natural compound (Figure 9), proved to be an inhibitor of this receptor and to decrease the proliferation of prostate cells when added to their culture.[50]

Can we think of treating tumours in the future using perfumes and flower extracts? The idea does not seem strange and fanciful any longer. Back in 2004 a paper reported on the effect of β-ionone and geraniol in reducing breast tumors, although the phenomenon had not been related to the action of olfactory receptors.

Another more recent piece of evidence strongly supports the idea that olfactory receptors do mediate cell proliferation. This time, the same Hanns Hatt group looked at the cells of skin and found still another olfactory receptor, also present in the nose and sensitive to a synthetic sandalwood odorant, Sandalore (Figure 28). When this smell was added to cells in culture, it stimulated their differentiation. Placed on a wound, it would accelerate its healing. Moreover, the scientists clearly demonstrated that these effects are mediated by an olfactory receptor as they were abolished when they silenced the relevant gene, thus blocking the synthesis of this receptor.[51]

Olfactory receptors everywhere

Recently we have witnessed a proliferation of reports which are finding olfactory and gustatory receptors in a variety of organs and tissues.

Receptors for bitter taste have been described in the trachea and other airway cells. These studies have also shown some therapeutic effects of bitter compounds in treating asthma and proposed that such substances could be used as a novel kind of drug.

The presence of taste receptors in the gastrointestinal apparatus has been linked to some sensory functions in the digestive system. In fact, even before these studies, receptors for neurotransmitters were known to be expressed in the gut leading to definition of this organ as a *second brain*.

Taste and olfactory receptors have been found in the heart, lungs, pancreas, kidney, and several areas of the brain. It seems that wherever we look for such receptors we are bound to find some. For this reason, some think their name *olfactory* or *gustatory* is inappropriate, although it is still true that the vast majority and diversity of olfactory receptors are expressed in the nasal cavity. Just as we have observed with OBPs which, both in vertebrates and in insects, include several members whose function is not associated with chemodetection, so also for olfactory receptors we should take a wider view and accept the fact that this is a large multigene family of proteins including members with unrelated functions.

OLFACTORY RECEPTORS IN INSECTS

While it was relatively easy to fish out olfactory receptors in different species of vertebrates, based on the first sequences identified in the rat, it took eight years to find the genes encoding olfactory receptors in insects. The reason is that these sequences are very different from those of vertebrates and the information accumulated so far was of little use. They were finally unveiled, but following a completely different approach.

The search was based on the partial genomic data published at that time for the fruit fly *Drosophila melanogaster*. In 1999 John Carlson, using a sophisticated informatic approach, managed to extract a series of sequences that were soon after confirmed to encode olfactory receptors. Almost at the same time, following a different approach, Leslie Vosshall and Richar Axel obtained similar results.[52]

Insect olfactory receptors still belong to the 7-TM family, but proved to be drastically different from those of vertebrates not only in their amino acid sequences. The first aspect which took scientists by surprise was the fact that they sit across the membrane *upside down*, that is with the C-terminus outside and the N-terminus inside the cell. An important consequence of this topology is that the region of the receptor assumed to be interacting with a G-protein, based on what is known for olfactory receptors in vertebrates, is found outside the cell. On the other hand, there is no evidence that G-proteins are involved in chemosignal transduction in insects.

So, how is the specific interaction with odorants conveyed to ion channels in order to produce an electrical signal? It has been suggested that the same receptors could act as ion channels. In fact, they associate with one member of olfactory receptors exceptionally well conserved across all insect orders and named Orco (olfactory receptor co-receptor). We know that the presence of Orco confers better sensitivity and specificity to all other olfactory receptors, suggesting that direct interactions between the two proteins should occur on the membrane.

Another unexpected finding was the small number, relative to those of vertebrates, of these receptors in insects. There are around 60 ORs in *Drosophila* and about the same number of gustatory receptors (GRs). We have already observed that in insects a distinction between olfaction and taste cannot be based on anatomical evidence. In fact there are olfactory sensilla not only on the antennae, but also on mouth organs, legs, and even wings in some species. Conversely, gustatory sensilla are also found in different parts of the body. Therefore, we can better talk about chemoreception to include both aspects, but that can be differentiated between detection of volatile molecules and contact chemosensing, which deals with non-volatile chemicals, such as sugars, salts, plant alkaloids, and long-chain hydrocarbons often present on the cuticle of insects. The number of ORs and GRs in other insects is variable, but within the same order of magnitude, with only few exceptions. The larger repertoire has been found so far in the jewel wasp *Nasonia vitripennis* with 300 genes encoding ORs and 58 encoding GRs, in both cases including 20–25 per cent of pseudogenes.

FROM RECEPTORS TO OLFACTORY IMAGES

An intricate bundle of electric wires

We have seen how the olfactory message encoded in structural parameters of the smell molecules is decoded and translated into electric impulses by the complex machinery of the olfactory neurons. These represent a physical interface between external environment and brain: a window of the brain onto the world of smell, but it could also be depicted as a skilled simultaneous interpreter, translating chemical words into electric signals, which are used by neurons to communicate with each other.

So now we can start to follow these electric signals along the complex wiring from the periphery to the brain. The connections are intricate and variable, continually adjusting to include new information coming in from the exterior world while establishing correlations with the data stored in the memory. Very little is known of the process and interactions of an olfactory signal on its way to the higher brain regions, until it is perceived as a conscious experience. Such investigations require contributions from several disciplines, not only biochemistry and molecular biology, neuroscience and electrophysiology, but also psychology to relate physiological data to emotions and behaviour, with computer science and mathematics to understand the logic underlying the network of neuronal connections and the language adopted by our brain for the efficient treatment of the data coming from the nose.

In turn, studying olfaction at its different levels and understanding the strategy adopted by our nervous system to process smell information can provide suggestions and models for assembling an artificial device capable of performing chemical analysis of the environment in real time, just as the nose does.

From nose to brain: the first steps

Let us now follow the pathways of electric signals generated by the primary olfactory neurons to the areas of the brain. The long tails of olfactory neurons, the axons, first cross a perforated bone, the ethmoid, located at the top of the nose and then enter the brain region. The destinations are the two 'olfactory bulbs', one on the left, one on the right. These are like a bunch of grapes, tiny beads, called glomeruli, clustered in a compact structure.

It is remarkable that all the neurons expressing the same olfactory receptor and therefore responding to the same smells, converge onto the same glomerulus. Imagine more than 1000 thin wires, coming from a relatively large area of the olfactory mucosa, which all end together in a tiny spot of the olfactory bulb. Then repeat this wiring for several hundreds of types of neurons and you get a most intricate bundle of strings all mixed up in apparently great chaos. Yet individual axons of each neuron find their way to the correct glomerulus without the need for traffic lights and road signs.

This is even more amazing when you think that olfactory neurons are being replaced all the time. Old neurons are discarded and new ones are generated from the stem cells present in the olfactory epithelium. These newborn neurons have to generate their axons and send them through the correct paths to establish the correct connections. It has been suggested that it is the same olfactory receptor proteins which guide the axons to the glomeruli. In fact, these receptors are also present in the axon, where certainly they can never be exposed to environmental stimuli.

But, it is still possible that such a sophisticated and efficient system can sometimes go wrong. The following anecdote was reported to me by a colleague while I was working in California and is the only case of this type I have heard of. A woman had her ethmoid displaced as a result of a car accident. In this process, all the axons of her olfactory neurons were severed and she completely lost her sense of smell. After a few weeks she started recovering and was able to smell again. However, this situation was much worse than being completely insensitive, because she was getting all the smells wrong. Imagine being in front of a lovely steak and smelling manure or getting a rotten fish flavour when you drink orange juice. I was told that in the end she did make a complete recovery and her ability to smell normally was restored. We still cannot tell whether her replacement of olfactory neurons eventually managed to find the right connections or if it was her brain that reprocessed the signals according to the information present in her memory.

A very efficient amplification

One of the functions of this complex machinery is to achieve a very high amplification of the peripheral signal, at least 1000 times, resulting from the addition of a large number of inputs coming from the individual neurons. But there is much more. The signal becomes much cleaner, so that even weak signals can be easily recognized.

When dealing with very weak electric signals we can certainly amplify them as much as we like, but at the same time we also amplify what is called the *background noise*, random signals generated by an instrument which is never perfect and can fire spontaneously even in the absence of any stimulus. In the same way, olfactory receptors suffer from this background noise, which would prevent signals that are too weak (at the same level as the noise) to be recognized. Adding together the signals coming from thousands of neurons has not only the effect of increasing the signal, but more significantly that of decreasing the noise. In fact, while an odorant compound stimulates all olfactory receptors of a certain type at the same time, spontaneous firing is random and these background signals coming from individual neurons cancel each other out in the process.

Similar strategies are applied to some spectroscopic techniques, where the same spectrum is recorded many times and all the data are collected together, thus increasing the real signals and decreasing the background noise typical of any electronic instrument.

Visualizing the connections

At this point the obvious question is how was it possible to reveal the complex network of connections discussed above? Again, it was thanks to recent techniques of molecular biology, which allow us to add, delete, modify, and connect genes, eventually obtaining a 'transgenic' organism exhibiting modified characteristics. Even today, such techniques are far from being easy and straightforward, but at the beginning of 1990s what Peter Mombaerts, at that time working with Richard Axel, managed to obtain was a great feat. He first introduced a gene encoding an enzyme and connected this gene to one of about 800 mouse olfactory receptors. He thus produced a transgenic mouse, in which, wherever that particular olfactory receptor appeared, the enzyme was also present.[53] To visualize the enzyme, and consequently that specific receptor, it was enough to add to a section of the olfactory epithelium a chemical substance which, oxidized by the enzyme, became blue and thus visible. In this way, he could see long very thin blue threads beginning on the surface of the olfactory epithelium and travelling towards the brain, across the perforated ethmoid bone, to converge on a dark spot in the olfactory bulb (Figure 29). A more impressive result was obtained by linking the gene for a fluorescent protein (GFP: green fluorescent protein) to that encoding an olfactory receptor. In such a case, the network of connections could be visualized even in a fresh tissue by merely irradiating the sample with UV light.

In insects, despite the drastically different architecture of the antenna and the sensilla with respect to the olfactory mucosa of vertebrates, the connections follow more or less the same patterns. A large part of the forebrain of insects is occupied by the antennal lobes, the equivalent of the olfactory bulbs, constituted by an assemblage of glomeruli. Using the fruit fly *Drosophila melanogaster* Richard Axel and his group reproduced very similar patterns of connections as those shown by Peter Mombaerts in the mouse. Based on a large number of data, a rule was established, both for vertebrates and for insects, that each olfactory neuron expresses only one type of olfactory receptor and that all neurons showing the same receptor and therefore responding to the same smell, converge to the same glomerulus.

An olfactory image in the olfactory bulb

Following the above assumption, which has been generally verified to a large extent, we can use our imagination to build a map of olfactory responses at the level of the olfactory bulb and visualize a sort of image for each type of smell. How complex is this map? It is not too

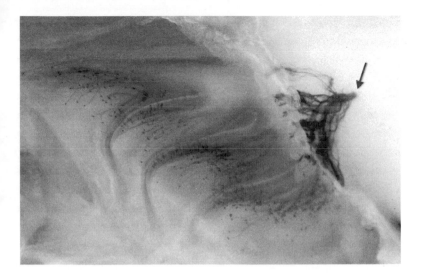

Figure 29. Visualizing the neural connections from the olfactory mucosa to the olfactory bulbs. Peter Mombaerts produced transgenic mice, which carried a gene coding for an enzyme coupled to only one of the nearly 1000 olfactory receptors. Whenever this particular receptor was expressed, the enzyme was also present and its activity could be detected with a chemical that turned blue. Thus, intricate connections could be visualized, from the tiny dots on the left marking the ends of olfactory neurons projecting into the external environment to a single point (indicated by the arrow) where all the signals converge. The wires (the long axons of the olfactory neurons) cross a perforated bone (a white area on the right) to enter the brain area, of which the olfactory bulbs represent the first station. (Modified from ref. 66 with permission).

difficult to answer such a question, thanks to a technique which allows cells to light-up following a stimulus. The entrance of calcium ions, a consequence of the opening of the corresponding ion channels, can be manifested using a probe that binds calcium and becomes fluorescent only in its bound form. Thus, we can stimulate the olfactory epithelium of a mouse or the antenna of an insect with a certain smell and look at the glomeruli of the olfactory bulb or the antennal lobe, having washed these organs with a drop of the probe. In this way, we obtain a real picture of how that particular smell is perceived. Giovanni Galizia, then at the University of Berlin, was the first to apply this technique to olfaction and obtained stunning images of the honey bee's antennal lobes responding to different smells.[54]

Even with hundreds of glomeruli, as in the mouse, we are still very far from coping with the diversity and complexity of the smells present in the environment. Therefore we cannot assume that each odorant would stimulate a single receptor type and switch on a single glomerulus. In fact, what we do observe is a complex pattern of activation, where certain glomeruli respond more than others, while many remain unaffected. The smell we perceive is encoded in this complex map, with relative contributions from each glomerulus.

In Chapter 1 we described some considerations of how complex would be a system to monitor the chemical environment. We also compared the olfactory system with our two major sensory systems, vision and audition, in particular the perception of colours and hearing. Colour vision provides a most efficient way of discriminating between a large number of variations using only three sensors, while the perception of sounds is based on thousands of different receptors each specifically tuned to a single wavelength. The olfactory system takes features from both senses. It is equipped with a large number of receptors, enabling us to distinguish components of a complex mixture (as in grilled meat, wine, or perfume); but, to cope with the very large number of smell molecules present in nature, it also needs to measure the response intensity of each receptor and use their ratios to discriminate different types of smells, as in colour vision.

And what happens when the olfactory stimulus is not constituted by a single type of molecules, but, as is the general case, dozens or even hundreds of different compounds are present, each with its own contribution to our perception? As far as we know, each chemical will generate its own response map and all the maps will be superimposed upon one another to produce the final complex olfactory image. We are still able to perceive the aroma of a wine by experiencing the overall sensation, but if we want to start paying more detailed attention to what we are sniffing we can still detect single olfactory components, sometimes even individual chemical compounds or at least different olfactory qualities.

Of course this is feasible with the more complex olfactory system of mammals, while insects with only a few dozen olfactory receptor types perceive an olfactory image of the world that is more blurred and less detailed. But they have other resources. What is most important for an insect is to find the right mate and distinguish the female of its own species from those of related species, which emit pheromones of similar chemical structure. While the performance of the general olfactory system of an insect can be quite poor, the receptors dedicated to smelling their own specific pheromones are much more accurate and sensitive. At the level of the antennal lobe we can identify special glomeruli, sometimes larger than others, which specialize in detecting pheromones. In such cases, the coding is much more direct and each pheromone component only activates one or two glomeruli.

Next we will ask what is going on beyond the Pillars of Hercules— the pillars of the olfactory bulbs. Thus far, we know very little about this as few intrepid researchers have so far ventured into these troubled waters, where the pure olfactory messages mingle with other sensory inputs. When we talk about our sensations as humans, those first memories of smell wandering across the brain elicit emotions and generate behavioural responses depending to a variable extent on our personality, experience, and culture.

PART 4

AT THE EDGE OF IMAGINATION

10

SCIENCE OR MAGIC?

The Debate on Human Pheromones

A QUESTION WITH NO CLEAR ANSWER

Whenever you deliver a lecture on pheromones, or more generally on olfaction, to a lay audience, it is likely that someone is going to ask about human pheromones. After listening to the fascinating and almost magical aspects of insect chemical communication, curiosity prompts the listener to ask whether similar mechanisms exist among humans. This interest certainly conceals differing motivations, from the obvious hope of attracting the girl or boy you have been trying to approach for a long time, to the disturbing concern that your life could be conditioned by scents, and the extent to which your emotions and decisions are dependent on volatile chemicals produced by fellow humans. The discovery of volatile compounds able to affect people's choices would certainly find alarming applications in manipulating behaviours of humans, as we can do now with insects.

These questions obviously carry heavy economical consequences when one thinks of the unlimited possibilities for manipulating people's choices with appropriately scented publicity.

At present, however, we still cannot draw any conclusions on the topic of pheromonal communication among humans. Many of us are convinced that human pheromones do exist and are still waiting to be discovered, with some scientists having ventured far enough to propose likely chemical structures. Others are sceptical and dismiss the idea that during evolution we have preserved such primitive ways of communication which would by-pass our conscious perception. In fact, although smells are known to produce immediate emotions and to recall vivid memories, we always filter and analyse whatever information comes from our senses, in particular from the sense of smell, before making a decision and acting consequently. Or at least this is what we hope we do. When we pass a bakery and are engulfed by the captivating smell of cakes still hot from the oven, do we always resist the temptation? And are we strong enough to keep away from fried chips, even if we know that they are bad for our health?

We cannot overlook the fact that smells have profound effects on our lives. The various, complex, and irresistible flavours of the foods we like act like magnets. The fresh smell of a sea breeze or the resinous scent of a pine forest have the power to put us in a good mood. A woman wearing a nice perfume or a man smelling clean from a special aftershave can certainly be attractive. But, however powerful these messages are, we are still far from the concept of pheromones. It is true that smells and sometimes just the idea of pleasant scents are effectively used to advertise all sorts of products; it is true that we often feel weak when confronted by some intoxicating olfactory attractions, but we are still not yet dealing with pheromonal communication.

We defined pheromones in Chapters 5 and 6 as those chemical compounds produced by individuals of a species and eliciting codified behavioural responses in members of the same species. Therefore, pheromones have to be species-specific, both regarding their production and their effects. Moreover, they have to act in the same way on all individuals of the same species or a subset (sex, age, caste, etc.) without the mediation of culture, education, or other sensory modalities.

DIFFERENT APPROACHES TO A COMPLEX ISSUE

We can now try to dissect this problem from various differing angles, asking simple and direct questions, to which we shall search for the answers science has recently tried to provide. To find our way through the intricate maze looking for human pheromones, we shall adopt anatomical, chemical, physiological, and behavioural approaches, examining in each case facts and hints that might indicate or rule out the existence of human pheromones.

Broadcasting chemical signals

Pheromones act between individuals of the same species, therefore for pheromonal communication to be active we need to identify both the sources of the chemical stimuli and the organs to detect them in the same species. We can then examine anatomical structures which might be likely sites for pheromone synthesis and release.

Insects possess specialized glands for pheromone production. In some cases, like those producing sex pheromones in Lepidoptera and other orders, they are clearly visible, particularly when they are extruded in the process of *calling*, when the female advertises her presence to the males in the environment. However, there are other sources of pheromones, more concealed and difficult to detect, as in the legs, in the antennae, or in the reproductive organs. Honey bees produce a variety of pheromones in their mandibular glands and in general social insects present several glands in different parts of the body secreting different types of pheromones.

But as our attention is focused on humans, mammals can certainly provide better clues than insects. We have discussed the urinary pheromones of rodents, the salivary aphrodisiac steroids of pigs, the vaginal secretion of the hamster, and the lactating pheromones of rabbits. Recall also the pheromone glands of deer and musk rats, producing secretions mostly appreciated in perfumery, as well as the appeasing pheromones secreted by glands on the chin of the cat and other mammals. Where should we start to search for human pheromones? We certainly possess all these glands, but whether they secrete pheromones is still a question in need of experimental evidence. Although volatile chemicals are present in human urine, sweat, saliva, and other biological fluids, a pheromonal role has not yet been assessed for any of them.

To search for sites of the synthesis of pheromones, we have to follow other leads. We have observed that, both in insects and in vertebrates, including mammals, wherever a pheromone is produced, there is a binding protein to help solubilization and transport across aqueous biological fluids. The urinary proteins of rodents, the salivary proteins of pigs, and those present in horse sweat are some examples. These proteins are very similar or even identical to those present in the pheromone detecting organs (the nose of mammals or the antennae of insects): OBPs, both of insects and vertebrates, and CSPs of insects. We have also observed that the use of the same proteins in broadcasting and detecting such chemical signals is not surprising and provides economical management of the available resources.

Therefore an alternative way to detect the pheromones is to look for the presence of their binding proteins. The underlying idea is that if the organism invests energy to synthesize a specific protein, the chemical it binds and transports is likely to be an important tool for the survival of the individual or for the conservation of the species. What is the advantage in searching for proteins rather than more directly for organic compounds? Proteins are easier to spot than volatile chemicals. We can look at the genome to find out straight away if the corresponding genes encoding the proteins of our interest exist or not. Then, using the different tools of biochemistry and molecular biology we can search more specifically within organs and secretions.

There is also another advantage, making our observations more reliable. When we analyse the chemical composition of a biological fluid, we always come up with a relatively large number of different organic compounds. Which ones qualify as pheromone candidates? The final proof must come from behavioural observations, but a tentative list might be based on those compounds which we find attached to proteins like OBPs and CSPs.

If we want to follow this approach, our first question is: are there OBPs in humans? Searching the genome we find a single gene encoding a protein of the OBP family. This OBP has been detected in the nasal mucus, specifically in the olfactory region, but so far not in other biological tissues or fluids. A sub-group of OBPs has been specifically associated with pheromone transport in the saliva of pigs, the urine of rodents, and other biological fluids used by mammals in chemical communication. A gene encoding a similar protein in humans has been detected in the genome, but, owing to a specific mutation, this sequence fails to be transcribed into RNA and consequently translated into a protein. This mutation occurred earlier in evolution and is already found in Old World monkeys and apes.

Another lipocalin, very similar to OBPs, is synthesized in the prostate, but we still have no indication whether this protein could be a pheromone carrier. Reproductive organs are known to be likely sources of pheromones in mammals, as well as in insects. Could this prostate lipocalin act as a pheromone carrier? So far no one has yet reported any chemical ligand that may bind to this protein. What about female reproductive organs? Our knowledge is even more limited. We know that among primates there are several ritual and stereotypical behaviours involving inspection of the genital area and rubbing in urine together with other actions which appear to be aimed at marking with pheromones or investigating the nature of scent traces. Unique to primates is the use of fingers to transport secretions from the source to the site to be marked or to take such materials to the nose or the mouth for chemical analysis.

But we should not limit our search for pheromones to chemical communication between the sexes. We have been considering other types of messages in insects, particularly the social species. Pheromones warn of danger, indicate the source of food, and help to identify members of a foreign colony. Humans are also a social species, although not in the strict sense that we attribute to honey bees and ants. It is still possible that we exchange subliminal messages which convey beneficial information to our fellows.

Among human secretions, sweat is regarded as the richest source of volatile chemicals. In particular, axillary glands, quite apart from those of the genital areas, produce large amounts of sweat, which is loaded with all kinds of organic compounds. We find steroids coming from metabolism, fatty acids produced by an abundant bacterial flora, as well as smells that may betray our last lunch in an ethnic restaurant or our appreciation of garlic and onions.

Potential carrier proteins for such compounds have been identified in human sweat. One of them is a lipocalin, called apolipoprotein-D, not exactly an OBP, but quite similar, whose function is to transport lipids. In the human sweat this protein was found associated with a strong odorant, a short unsaturated fatty acid, 3-methyl-2-hexenoic acid, responsible for the typical odour of armpit sweat. Could this be a pheromone? If so, what sort of message could it carry? This odour might have been attractive among early human populations, but is now generally regarded as offensive and an indicator of poor hygiene—at least by most of us.

A well-known story relates that Napoleon used to send a message to his wife that he would be back home in a few days and asking her to refrain from taking a bath in the meantime. True or imaginary as the story might be, to some individuals the smell of the sweat of a lover is attractive. In the same way, the smell of manure, unpleasant by normal standards, can be agreeable when it reminds us of a rural landscape uncontaminated by exhaust fumes. Is it correct to compare the two examples? In the first case we are dealing with chemicals produced by individuals of the same species and we could consider the possibility that they may act as pheromones.

However, other characteristics of pheromones are missing. For example, sweat can create different sensations and reactions in different individuals, while pheromones are supposed to act in a more general way. Moreover, when someone finds the smell of sweat pleasant, this phenomenon is limited to the typical bouquet of a specific person, while the sweat coming from any other person, although similar in composition, would be perceived as repulsive. Therefore, we had better classify the smells coming from sweat as olfactory markers, which might in some cases recall good memories, and so become acceptable, like the smell of manure, rather than as pheromones.

Special organs and receptors for pheromones

Now, let us approach the problem from a different perspective and look at the anatomical structures dedicated to the detection of pheromones. Insects are endowed with special sensilla tuned to pheromones; and the signals originating from such sensilla are directed to a specific region of the antennal lobes, the first station of amplification and processing of chemical information. Mammals and other vertebrates also possess a special organ where pheromones are detected, the vomeronasal organ. This is a sort of second nose, which sends its neurons to a special *accessory* olfactory bulb, well separated from the main one, dedicated to processing general odours. Therefore, both in insects and in mammals, although with different anatomical structures, pheromonal messages use preferential channels by which they are detected, processed, and sent to the brain.

Now the question is: do humans possess a vomeronasal organ? It looks like a simple problem to solve, but even on this point there is no agreement. In fact, in the newborn infant a structure recognizable as a vomeronasal organ can be observed. It consists of two very small dead-end channels with their openings at the base of the nostrils and projecting towards the nasal septum. They are very small and you need a microscope to see them. As we grow and develop, these channels atrophy and disappear in the majority of adults. In those few individuals where such structures can still be observed during adult life, they are not innervated and therefore are not connected to the brain. As for the second structure, accessory olfactory bulbs, the answer is easier: they do not exist in humans.

Such observations strongly suggest that the vomeronasal organ is mostly vestigial in humans. The disappearance of a functional vomeronasal organ occurred earlier in evolution, as this structure is absent not only in our closest ape relatives, such as gorillas, chimpanzees, and orang-utangs, but also in Old World primates. Can we therefore conclude that pheromonal communication was abandoned as we evolved into higher primates and eventually into human beings? The scenario is rather more complex. There are other tiny olfactory organs, located in the nasal cavity, such as the Gruenenberg ganglion and the septal organ of Masera, which may also be involved in pheromone detection.

Let us first have a look at other tools associated with pheromone detection and focus our attention on proteins. This time we are not looking at binding proteins, involved in the delivery of pheromonal messages, but at receptors, necessary for their detection. We might remember that two specific families of olfactory receptors are expressed in the vomeronasal organ, named V1R and V2R. These receptors still belong to the large superfamily of seven-transmembrane receptor proteins, but are markedly different from olfactory receptors. In particular, we may recall that V2Rs are endowed with a very large extracellular domain, probably the site of binding for pheromones of protein nature.

A search through the human genome informs us that we do not have any V2R, but around 200 genes encoding receptors of V1R type can be found. Most of these, however, are non-functional, and we are left with only five complete genes. So where are these genes expressed? At least one has been detected in the olfactory epithelium, but also in other tissues, suggesting to some that we can still communicate through pheromones, using our main olfactory organ. This idea is supported by some evidence that other mammals can detect pheromones with their olfactory organ. But for those who cannot accept the idea of human pheromones the migration of this gene to the olfactory region just means that the encoded receptor may have been adapted to different functions.[55]

Meaningful smells or automatic switches?

We have so far been trying to collect evidence supporting or disproving the idea of human communication through pheromones by looking at secretory glands where such chemical messages could be produced and also perception systems dedicated to their detection. But what about the molecules of pheromones themselves? Can we form hypotheses on what they might look like? Are there particular chemical classes that we should focus on to search for such candidates?

When we consider the variety of insect pheromones, we realize that any chemical could be adopted for a pheromonal function. Even among vertebrate pheromones, in particular those of mammals, we find long chain fatty acids, aromatic compounds, steroids, macrocyclic structures, sulfur derivatives, and many other classes of organic compounds. Within this variety, steroids have been the object of special attention, probably due to the fact that some are excreted as by-products of metabolism in a sex-dependent fashion. Androstenone, a molecule we have met on more than one occasion, together with its relative, androstadienone, are produced from dihydrotestosterone and testosterone, respectively, by loss of a water molecule. We have already observed how such chemical modification converts odourless steroids into more volatile compounds, endowed with potent characteristic odours. In fact, these chemicals have been demonstrated to have pheromonal activity in pigs and might also be active in other mammals.

Although androstenone smells repulsive to humans (at least to the half of the population who are able to smell it), one of its corresponding alcohols, androstenol, is endowed with a pleasant musky scent. We might recall that the name of this olfactory note comes from the secretions of musk deer, musk rat, and other species, which produce and use as pheromones chemicals structurally unrelated to steroids (macrocyclic ketones and lactones), but smelling very similar to androstenol. So, all these elements link pheromones, sex, steroids, and musky scent to one another and perhaps suggest that human pheromones might fall among such odorants. In fact, the thread to follow is rather weak and not fully convincing. To add confusion,

there are reports, not confirmed by subsequent research, that androstenone is perceived differently by human males and females, and that
perception of the musk smell changes during the menstrual cycle.

Not only sex

When mentioning human pheromones our imagination generally
heads towards sex pheromones and the potential use of these to
attract individuals of the opposite sex. This mechanism works well
in insects where, besides acting as aphrodisiacs, pheromones perform
the important function of labelling females with tags identifying the
species and thus preventing unproductive mating.

We certainly do not need information of this type to recognize a
member of our own species, and in this respect can do without sex
pheromones. We can therefore shift our attention to other types of
pheromone, if we still want to pursue the search for such messengers
in humans.

A few years ago two independent papers reported experiments
suggesting that humans can communicate their stress levels to other
individuals through the smell of sweat. They addressed two different
situations producing high stress, the first generated in students waiting
for an academic examination, the other in individuals jumping from a
plane during the single minute of free fall before opening the para-
chute. The sweat collected in such situations was presented to a
number of subjects while their brains were scanned for activity. As
controls, sweat produced during physical exercise was used. In both
studies, the sweat during stress stimulated areas of the brain related to
emotion, unlike the control samples. Are we then in the presence of
pheromonal communication? It is difficult to say: it could still be a
learned behaviour, but we cannot rule out the possibility of messages
unconsciously released and perceived. When the molecules responsible for such effects are identified, they will certainly provide important information and tools to clarify this issue.

We can complete this rather inconclusive meander across secretory
glands, putative receptors, and potential pheromones with perhaps the

most likely situation in which we might come across human phero-monal communication, if such a phenomenon exists at all. We have observed that the vomeronasal organ is vestigial in humans, but can still be witnessed in newborns. Perhaps this is the stage where we need to be guided by a scent trace to find what we most need at that age: milk.

Benoist Schaal, who discovered the lactating pheromone in rabbits and clearly demonstrated how the rabbit young follow an innate attraction to this chemical cue, also investigated humans during their first days of life. Working with humans, particularly with new-borns, presents a series of ethical problems, which limit the types of experiments that can be performed. Neverthless, Schaal found that newborns can smell and are attracted to scents originating in the nipple area. These are very complex in composition, containing secre-tions of glands situated in the nipple areola, but also smells coming from milk or colostrum, to say nothing of aromatic components contained in creams and lotions applied by the mother to the area. So far, none of the compounds examined can qualify as a putative pheromone, but if a human pheromone does exist, this is probably the area where we can expect such a discovery.

We can now try to put together the scattered and incomplete data available in the attempt to solve this puzzle without any idea of what the final picture will look like. We have searched for sites of the production of pheromones and suggested sweat glands and secretory glands around the nipples as likely candidates, but failed to detect specific chemicals produced in these sites. We have also failed to find OBPs, which might betray the presence of pheromones to which they would act as carriers, although we should still consider the possibility that other binding proteins could perform this task in humans. The absence of a functioning vomeronasal receptor might rule out the hypothesis of pheromonal communication in humans, but migration of our single vomeronasal receptor to the olfactory area has opened up the possibility that we could detect pheromones through our main olfactory organ. Moreover, a vomeronasal organ might be acting in newborns with mechanisms still to be discovered. Similarly, a chemical approach has failed to provide clear proof either in favour of or against the idea that some kinds of smells might act as pheromones in humans.

Perhaps the most convincing argument against the existence of human pheromones, at least sex pheromones, comes from cultural and evolutionary factors. The strongest pressure to make sex pheromones which may represent robust and specific signals, is found in insects, where, due to the enormous variety of species, with many of them closely related and sharing the same habitat, the danger of mistaking the correct partner could be very high.

In mammals, rather than just identifying the species, sex pheromones may advertise the fitness of the male (more often than the female, unlike amongst insects) and a source of better genes for the female. In this respect, we can observe that with birds, in most of which the occurrence of pheromonal communication does not exist, the choice of the partner is based on visual and auditory clues. Vision also plays an important role in primates and notably in humans when a partner has to be selected. While we do not need to smell a potential partner to make sure he or she belongs to our species, other elements, such as appearance, culture, and social position, become more relevant in ensuring our offspring will grow up in a safe and stimulating environment, optimal for the development of physical and cultural qualities.

The tentative conclusion of all this is that it is most likely that pheromonal communication does not operate in humans, yet still leaving a little window open for chemicals mediating the relationship between mother and newborn.[56]

Despite such poor evidence in favour of human pheromones, we are continuously bombarded with advertisements promising lotions and perfumes endowed with the magic properties of attraction for individuals of the opposite sex, often supported by pseudo-scientific publications in obscure journals. It is perhaps not too surprising if many people are persuaded to buy such products, given the faith some of us have in, for example, card reading or some so-called homeopathic treatments.

DIGITAL OLFACTION
Detecting and Reproducing Smells

SCIENCE FICTION OR THE NEXT APPLICATION?

Can you imagine switching on your television or searching the internet on your computer or your telephone and, besides sounds and images, receiving olfactory information? Not just descriptions of smells, but real smells you can sniff, molecules wafting up your nose. Just think of a nature documentary enriched with the scents of flowers, the breeze of the sea, and the fresh air of a prairie or a forest. Or a gastronomy programme in which you can smell the dishes being prepared and sniff the bouquet of a wine bottle as the cork is popped. It is hard to find anybody who has not dreamed of such possibilities, wondering at the same time why we are so advanced in transferring sounds, words, images, and colours through the air with extraordinary accuracy and fidelity, but we are still not able to do the same with smells.

Is it just a matter of technology, which is not advanced enough to perform such tasks?

In a sense, this is partly true. The olfactory system is too extremely complex to reproduce with the currently available tools. But, more significantly, our understanding of the chemical language, which is the basis for translating chemical messages into perceived smells, is still too poor and incomplete. It is a bit like writing a software program for translating text into another language: first it is necessary to be able to read and understand the language we want to translate with precision.

The main point when we come to work on smells is that we get very demanding and do not accept anything which is not perfect. To keep the analogy with the translation of a text, reproducing smells is something like translating a poem. A poem is not a list of data or an assemblage of information which you can find in the manual of a washing machine. The translation has to convey the same emotions, when put into another language, which the author wanted to communicate in the original. We are ready to accept faded or slightly blurred images if the story of a film is captivating enough. We can even put up with an imperfect reproduction of sounds (you may remember the excruciating metallic ringtones of mobile phones only a few years ago, pathetically trying to reproduce pieces of classical music), but we seem to require high fidelity when it comes to smell. Smells recall memories and elicit emotions in a more direct way than images and sounds. If olfactory signals do not match exactly those we have stored in our memory, they fail to produce the desired effect and our reasoning cannot come to our aid as it does with pictures and sounds.

Let's take a look at the tools and information needed to bring smells and olfaction into the digital world and some of the many attempts, so far unsuccessful, to construct an electronic nose or to reproduce fragrances and flavours in a computer or a telephone. But before venturing into more technical matters, let us ask ourselves to what extent electronic devices are capable of analysing, transmitting, and reproducing smells so that they might improve our experiences and our life in general.

Are we ready to smell anything?

It is not always accepted that a message containing visual and acoustic elements could be further improved by the addition of olfactory experience. In fact, smells can be more aggressive than other sensory stimuli and it might be difficult to get rid of those we do not want. If we hear an unpleasant sound or catch a glimpse of a disturbing image, we can switch off our device and immediately delete such intrusions at the touch of a button. Or we can turn our eyes from something we do not like to see or to some extent cover our ears if we do not wish to listen. It is not so easy to do the same with smells. Some odours can be repulsive and cause nausea, headache, or irritation in sensitive subjects. Such reactions cannot be eliminated merely by removing the source of the odour. Molecules linger in the air and right inside our nostrils and even when they have gone we continue to have a feeling of smelling something.

But all these problems and limitations are no obstacles to the persistent urge to make our dream come true. The possibility of sending olfactory messages just as we habitually do with images and sounds has fed the imagination for a long time and has become a dream of many scientists. In our virtual world it is common to receive virtual flowers for your birthday, just the colourful image of a bouquet. The addition of the appropriate scent would greatly contribute to making the e-card more realistic and emotionally effective. A restaurant could include in the menu, besides attractive pictures of the dishes, also their flavours to anticipate what they smell like. We do not need to list the innumerable applications in the field of advertisements, where the evocative power of smells would make promotional messages more compelling even beyond the limits of ethical behaviour.

Such widespread interest and strong expectations have been the fertile ground for improbable devices sprouting like mushrooms and promising wonders with a host of new words emerging to catch the attention and stimulate the imagination. *Smell-O-vision* projecting scented films, *O-Phones* claiming to send smells through mobile phones. *Green Aria* was a live opera enriched with smells; *Sound and Perfume goggles* are connected to smart phones and emit smells when someone you know is around; *Smell Screens* are equipped with electric fans spreading the flavours of the foods they advertise.

These are only some of the many attempts to harness smell which keep appearing on the internet. But on 1 April 2013, the long awaited news was released! Finally it was possible to send smelly messages through *Google Nose*, a new application promising wonders. Not everybody paid enough attention to the date of the release and many sent enquiries for more information. We are still, however, very far from being able to attach smells to our e-mails, but all these facts testify to the high expectations and interest in this field.

A BIT OF HISTORY

We do not need to go far back in time to trace the ideas and the attempts to build an artificial nose. The name of course is too pretentious and suggests capabilities far above the simple and basic performance of current devices. However, names like *artificial nose, electronic nose*, or more simply *e-nose* have found their way into common speech and indicate the final goal rather than the current situation.

Arguably the first documentation of a device built with the idea of discriminating smells is a paper published in *Nature* in 1982 by Krishna Persaud and George Dodd.[57] At that time some simple gas sensors were commercially available and used to detect house gas leaks. Such sensors are nothing more than pellets of metal oxides, whose electrical resistance changes when in contact with various gases. They respond to many chemicals in the gas phase, but with very poor specificity. However, using three different types of such sensors, Krishna was able to assemble an instrument capable of identifying several different chemicals in the gas phase.

It was the first time that the possibility of discriminating smells was experimentally put to work in a functioning device, however basic. The main idea was that of the *combinatorial code*, which was later verified in the biological olfactory system. In simple terms, we do not need a specific sensor for each type of smell, but we can use detectors with broad overlapping responses to recognize a large number of stimuli, provided we can measure the signals produced by the sensors with sufficient accuracy. The first electronic nose assembled by Krishna was similar in principle to our colour vision. In both cases we deal with three types of sensors, in both cases their selectivity is rather poor, but the responses can be measured with good accuracy.

While the idea of the combinatorial code remains the necessary basis for the design of any device for odour discrimination, the metal oxide sensors could not offer the versatility required to cope with the large variety of smells which the human nose can detect. However, that seminal paper remains as a landmark in the history of artificial noses, demonstrating for the first time the feasibility of designing such instruments and indicating the path to follow.

A versatile family of gas sensors

At about the same time that I was studying our newly discovered proteins, which would later be called OBPs, and trying to make sense of their function in olfaction, I was already nursing the idea that these same proteins could be used as sensors for smells in an artificial device. But we knew that proteins are very delicate molecules and should be treated with great care to avoid changes of shape which would result in modification or loss of their activity. Therefore, the idea of using OBPs was quickly dismissed at that time as too adventurous and not practical. Curiously enough, the most recent research on artificial devices for smell monitoring is focused on OBPs as the most promising sensing elements.

After putting aside proteins, the attention turned to polymers, also thanks to a fortunate coincidence. You do not need a brilliant imagination to see that polymers are large and versatile enough to encapsulate different organic compounds and thus could be tailor-synthesized for different needs. The main problem was how to obtain a measurable signal from such interactions. The gas sensors used by Krishna were conducting elements, and their electrical resistance could be monitored. But organic polymers are insulating materials, so good that they are widely used in protecting and insulating electric cables and in similar applications. We needed *conductive* polymers, an apparent contradiction. But such materials did already exist, although they were not as popular as they are now. At that time most of the attention was on polymers obtained from acetylene, the gas used to illuminate streets before the invention of the electric bulb and adopted until recent times by speleologists for the long autonomy it provided.

Acetylene is a very simple molecule of only two carbon atoms connected to each other by a triple bond. This bond can be partly opened to generate a long chain of carbons linked to one another by alternating single and double bonds. Such polymers, when opportunely *doped* with ions, could conduct electricity. Besides the fact that these long molecules were highly unstable, for our purposes a single conducting polymer, like polyacetylene, could at most represent only one of the several sensors needed for an electronic nose.

It was during a lecture delivered in my department by a visiting American scientist, that I learned about another class of conducting polymers, long chains of repeating units of pyrrole. These polymers at the time did not receive much attention because of the problems they presented. First, their electrical resistance was too high to be used as substitutes for copper wires, then their conductivity was not stable, being affected by vapours of ammonia and other gases. These *disadvantages* were exactly the characteristics required for versatile gas sensors. In addition, the ring of pyrrole could be modified by attaching all sorts of chemical groups and chains, thus altering the specificity of response to different gases.

Soon after, Krishna spent a long period in my lab to purify and characterize OBPs, as I have recounted in Chapter 8, and we combined our efforts in this research too. The idea was to prepare several derivatives of pyrrole that would be used for conducting polymers with different overlapping specificities.

In a few months the first prototype was assembled. It made use of 20 different sensors and was capable of discriminating compounds of the same chemical class (such as alcohols or ketones or amines) differing by only one or two carbons. The software, entirely written by Krishna, was running on a Commodore 64, probably one of the first personal computers to be used in the home, endowed with a RAM of 64 kB and using a cassette recorder to store data.

These results aroused immediate and widespread interest and things started moving at a fast pace. Krishna moved to the University of Manchester where he further developed an electronic nose, which soon became commercialized. A prototype was exhibited at the Science Museum in London and a compact version was installed on board the MIR space station where it collected data for several years.[58]

Simple devices with some applications

Many labs started working on electronic noses, most of them using sets of conducting polymers as detecting elements, sometimes combined with the metal oxide gas sensors. Several companies were founded and some are still active. The name *artificial* or *electronic nose* was too appealing to be replaced by more realistic ones.

Certainly such devices present interesting features and advantages. They can generally perform gas analysis in real time and without the need to separate the components of a mixture. In such characteristics they resemble a biological chemical detector and have found useful applications whenever a preliminary screening of a large number of samples is important for later selecting a few to be subjected to more accurate and specific analysis. They are also used in environmental monitoring, where they can send alarm messages in real time to warn that something in the parameters is changing. In fact, instead of performing an olfactory analysis, our currently available devices can monitor the atmosphere above a certain sample and detect any variation.

These current devices are still very crude, but have already found applications (in combination with chemical analysis and sensory evaluation) in the food industry to ensure that the aroma of the products is constant; and in environmental monitoring of the quality of air and water, as well as tentative uses in some medical preliminary screening.[58] In fact, several diseases, including cancer, are often accompanied by production of volatile compounds which animals can easily detect.

Many reports have shown that dogs can smell cancer in the early stages and a cat became famous a few years ago for *predicting* death. This cat was living in a hospital and used to visit those patients who almost invariably died on the following day.[59] The use of animals, like that of an artificial nose, can provide preliminary indications of patients needing more specific attention and further analyses.

So far we have been talking about instruments performing chemical analysis of gases with some practical applications, but we are still very far from an artificial system even vaguely resembling the complex performance of a biological nose. Perhaps at this point we can return to more basic questions and discuss what an artificial nose should look like, what kind of performance we expect from such an instrument, and what we need in terms of hardware (sensors) and software (information) to design and build an electronic nose.

WHAT IS AN ARTIFICIAL NOSE?

Before designing an artificial nose, we need to decide what we want to make. The question is not trivial when we consider that in talking about a *nose* we do not simply refer to the anatomical protuberance on our faces, but to a complex system for detecting and recognizing smells, a large part of which is in the brain. The molecules confronting our noses eventually produce verbal descriptions, emotions, and behavioural responses. To a large extent these external consequences to the detection of a smell are mediated by our experience, memory, mood, and personality, and are therefore subjective.

Hardware and software

Any instrument for environmental monitoring, including an electronic nose, is the combination of hardware, represented by sensors, performing some sort of measurement, and software, which can process the analytical data and interpret their significance according to some guidelines. In the case of an electronic nose, the hardware is an array of chemical sensors, capable of interacting with the smell molecules and reacting to those parameters which are relevant for our olfaction.

Chemical sensors

Can we use any kind of chemical sensor? It depends on how we want our instrument to resemble a biological nose. We have described two types of sensors, the metal oxides and the conducting polymers. Both can discriminate between different chemicals, but do they use the same criterion adopted by the nose? In Chapter 3 we discussed which molecular parameters could best correlate with the different odour types. We noted that, to a large extent, the shape of a molecule and its size are more important for smell than the functional groups. We also observed among other examples that alcohols of different shape smell different, such as the grassy-smelling 3-hexenol and the mushroom odorant 1-octen-3-ol, but their olfactory characteristics do not change much when the alcohol group is replaced by an aldehyde or a ketone. This is the type of information we need in our choice of sensors. We want sensing elements which respond in a similar way to 3-hexenol and 3-hexenal, but which can discriminate between 3-hexenol and 1-octen-3-ol. If our chemical sensors cannot distinguish between these two alcohols, which definitely smell different, no software will be able to do the trick.

Evidently, even to design the hardware and to choose the sensors for our electronic nose we require at least a basic knowledge of how our biological nose works. We can say that an artificial device reproducing the functioning of our nose should use the same type of language, although not necessarily the same alphabet. So far, conducting polymers are the best suitable sensors with such characteristics, although their performance is still very primitive compared with the proteins of our noses.

Biosensors for an electronic nose

Then, why not use olfactory receptors? As far as we know, membrane proteins, like many types of receptor, are delicate entities and need the complex environment of the membrane to maintain their special folding, which is essential for preserving their binding properties. At the present state of our technology it is inconceivable to incorporate olfactory receptors into electronic circuits and expect that they can still bind and recognize odours.

But there is another class of proteins involved in olfaction and contributing to the discrimination of different smelling compounds. We have extensively described the structure and the characteristics of OBPs. In particular, we have observed how the compact structure of these proteins makes them refractory to thermal denaturation and harsh environmental conditions. Besides, their synthesis is simple and cheap, allowing for mass production of biosensors based on OBPs. With respect to the other sensors described here, proteins present unique flexibility for modification, through mutations at specific sites, in order to meet with special binding requirements. This possibility, which has been experimentally verified with several OBPs, is based on our detailed information of the three-dimensional structure of a large number of these proteins, as well as on computational tools able to predict with satisfactory reliability the effect of specific mutations on the binding properties of a protein.

Currently, the use of OBPs as biosensing elements represents the cutting edge of research aimed at modelling olfaction with electronic devices. The weak point is still the transduction process. How do we obtain an electric signal from the uneventful binding of an odorant molecule to a protein? Some successful attempts have been reported, although we are still far from the reliability we need for a commercial device. OBPs have been incorporated into *bio*-transistors, which are able to respond with an electric signal when an odorant molecule is captured by the protein. Other approaches have addressed the optical properties of OBPs and their changes in the presence of smell molecules. The technology required for both approaches is quite advanced and these fields are progressing fast.

How many sensors?

To build an artificial nose we need a very large number of sensors, because our olfactory language is based on a large number of smells, each with its own character. In a mixture containing cinnamon, cloves, coconut, lemon, and many other types of smell we can detect and recognize each of them. Although smells are recognized using a combinatorial code, it is also true that the elements of this code are many and each is perceived by the nose in a different way. Smells do mix, but only to some extent, as we observed more than once, and the olfactory combinatorial code is in no way similar to the colour code we use in vision. We can easily produce yellow by mixing green and red lights or violet from red and blue, but we will never be able to reproduce the scent of roses, to give an example, by mixing mint and cheese or lemon and pepper, or other common scents.

Our sense of smell is based on hundreds of different receptors and they are necessary for giving our olfactory experiences the richness and diversity which enable us to appreciate wild strawberries better than those cultivated, or to recognize the special flavour of our grandmother's cake, similar, yet nevertheless different from the product you buy at the local shop. The more sensors we use the better can we describe our sample in terms of smell. Of course, we can think of a wide range of instruments, from the very basic ones, equipped with a limited number of sensors and used only for some specific tasks, to a complex general purpose analyser, which might approach the definition of an artificial nose.

Relatively simple instruments, based on a few dozen sensors, are already available and have found practical applications, although being far from offering the performance of an artificial nose. What can we ask from such instruments? At a better level of sophistication than the present one, an instrument equipped with sensors reproducing, yet with a limited number of elements, the basic responses of our biological system, could evaluate the flavour of a food, using the same parameters employed in our sensory analysis and identify which samples are similar in their smell and which are different. Such an analytical tool would prove invaluable when performing quality controls in the food industry.

Currently, the evaluation of the aromatic properties of foods is performed by panels of experts who smell and taste all the samples and for each of them provide a series of scores for all *descriptors* previously identified, which contribute to identifying the aroma of a specific product. The list of descriptors is often quite long with several dozen terms and the work of the experts is time consuming and expensive.

In such cases, an artificial nose would be able to verify that the organoleptic quality of a product remains constant, as long as the profile generated by the sensors of the instrument matches a reference profile obtained with a sample that has been rated as *good* by the panel of experts. Therefore, we will always need human subjects to judge the quality of a product in the first place, but then the electronic nose can take over and confirm that the *quality* remains constant. When a sample produces a different response, what the instrument can tell us is that something has changed and a detailed investigation is required.

From chemical sensors to a sniffing device

We have assumed, without explicitly stating the question, that we want to build an electronic *human* nose. We want something as similar as possible to our nose, rather than the nose of a mouse or the antenna of an insect. In practice, we need a sort of translating machine capable of performing chemical analysis on a mixture of volatile molecules and providing an answer in terms of smell descriptors. As these descriptors belong to our perception of smell, we should teach this machine our olfactory code to make sure that the particular bouquet of volatile compounds which we perceive as, for example, roses should also be labelled with the same term by an electronic instrument.

All this should be the task of sophisticated software capable of describing analytical data with the words used by a perfumer or a food taster. But how far can we go? Certainly an instrument should measure smells according to standard rules, independent of individual factors. Describing and categorizing smell is equally complicated for an electronic nose as it is for a panel of human judges. To standardize responses we have devised the tools of sensory analysis, which, using a series of descriptors, help the judges to translate a perceptual sensation into a fixed framework of categories and types of smell. An electronic device might be equipped with a software program using a similar strategy and relying on a database of descriptors for every shade of smell, built on the experience of trained judges mainly working in the perfumery and the food industries.

But, an artificial nose, however sophisticated, will never tell us if an odour is pleasant or not and whether a certain wine is better than another. When the response is affected by emotional components, it becomes too personal to be measured by an instrument. This concept is obvious too for other sensory modalities. We can reproduce a painting using a camera and make all kinds of analyses, from the colours used to the perspective and the composition of the scene, but no instrument can tell us whether a Michelangelo is better than the face your five-year-old son has sketched, or how to compare a Rembrandt with a real life photo. Similarly with music, we can dissect all the notes of a symphony and analyse the contributions of each instrument, but we can never understand, using measuring instruments, why that particular concerto makes you happy and relaxed, while another kind of music is tiring and boring for you, and the opposite happens with another listener.

We are well aware of such limitations in scientific instruments when dealing with other senses, but the same concept is less obvious with smell. The reason is most likely linked to the unique efficacy and immediacy of olfaction to stimulate our emotional areas and the strong links well established and stored in our memory with past experiences. Nevertheless, this will certainly be the limiting factor for artificial noses. We can imagine that at its optimum performance it might provide the kind of responses we now get from a panel of perfumers or food tasters.

Future targets

In addition to a better knowledge of our biological nose and the complexity it conceals, one of the main elements of concern is sensitivity. Current devices based on conducting polymers offer operations at concentrations of orders of magnitude higher that those detected by our noses, let alone the nose of a mouse or the antenna of an insect. To overcome this difficulty, some instruments include a concentration step before the sample is analysed. However in this way, a basic requirement of an artificial nose is missing, the capacity of sensing in real time, and the device resembles more a laboratory instrument than a sensor. Again, we could learn a lesson from the biological nose. We have seen how olfactory neurons send their signals to the olfactory bulbs, all those responding to the same type of smell arriving at the same location. The result of this convergence is a tremendous amplification of the signal with a drastic reduction of the *noise* (the spontaneous firing of neurons). This strategy has already been successfully applied in several analytical instruments and could be easily adapted to improve the sensitivity of electronic noses.

We can summarize this challenging exploration into the possibilities of digitalizing smells with cautious but optimistic predictions. Instruments attempting to detect smells using the same approach of our nose already exist, although their perfomance is still basic and limited. We have a path to follow and we have identified the tools we need. Of basic importance is a more detailed knowledge of our biological olfactory system in order to identify the elements of the chemical language understood by our nose. At the same time, we need to develop better sensors, perhaps using novel strategies and materials. In this respect, the use of proteins as specific elements to finely recognize the various molecular shapes of odorants looks more and more feasible.

CONCLUSION

We started our adventure by paying closer attention to the smells around us, being guided by our nose to pleasant scents and attractive foods, and away from bad smells which forewarn us of potentially dangerous situations. We have learned how different smells are encrypted in the molecular structures of volatile compounds and how they represent the alphabet and the words of a complex language. This language, that we humans are beginning to decipher, is spoken by most animals to advertise their presence to individuals of the same species, to send warning messages, to exchange information about food sources, but also to eavesdrop and deceive and eventually exploit.

The study of chemistry has supplied us with tools to articulate the words of this foreign language, while biochemistry and molecular biology, by unveiling receptors and neural connections, can explain to us how the varieties of chemical messages are interpreted by each species to eventually elicit their behavioural responses. In humans we are just starting to understand how smells are so powerful and direct in eliciting emotions and recalling long lost memories.

Having unveiled these secrets, carefully guarded for a long time in the deep recesses of our noses, and understanding how molecules in the environment can produce olfactory images in our brain, a question remains: have we lost that magical, elusive savour associated with smell? Do we no longer associate pleasant smells with poetry and romance?

Certainly we may still by-pass the scientific facts and let a pleasant smell guide us in the realm of emotions and imagination. But sometimes the fascination of the unknown dissolves when we explore a new place, when we learn a new language, even when we get to know our partner better and better. The superficial attraction of mystery and magic leaves room for knowledge, which is the ultimate interest and pleasure for human beings. The compelling attraction of the unknown prompts us to engage in adventurous discoveries, and the beauty of discovery is that this endeavour never ends. The opening of a door leads other doors to be opened and entire unsuspected realms to be discovered, while our surprise and excitement is renewed each time.

Understanding olfaction and the mechanisms of perception, the structure of receptor proteins, and the intricate neural connections from the periphery to the brain, all helps us to put chemical communication into a wider context. Ants and bees maintain their well organized societies thanks to an invisible network of volatile molecules connecting each individual to the others. These are flexible and ever-changing connections. Yet still it is a very robust and stable system ensuring that the rules of the community are always followed so that the life of the nest proceeds efficiently. Slime moulds are also made of individual units (in these cases they are cells) which however have the capacity to assemble into an organized organism. The neurons of our brain, relatively simple when observed individually, are capable of complex higher functions when put together and interconnected. The cells of our body have been assigned different tasks and have differentiated accordingly.

In this sense our body is a superorganism with different cells performing distinct tasks for the benefit of the organ, ready to commit mass suicide when the time has come for them to be replaced. The signals regulating the perfect functioning of this clockwork are molecules, small molecules as messengers, and proteins as receivers of the information. Ants and honey bees share several aspects of the cells of our body, but can we talk of a superorganism? Certainly, individuality is lost to a large extent and these insects are ready to die *en masse* to save the hive or the nest. And if we accept that the idea of a super-organism can be applied to social insects, what about humans?

We are also a social species, although in a rather different sense. We live in communities organized in such a way that the society as a whole can be independent and self-supporting. We are connected by a net-work of relationships, not smells of course, but language, empathy, common interests, also economic exchanges. But the messages we exchange within our human community are not as strong and com-pelling as pheromones. All of them, including olfactory messages, go through the filter of reason. Our choices are conscious and always the result of motivated decisions. Or are they? Unfortunately, this is not always the case and, even in the absence of pheromones, we sometimes follow orders to act against ourselves and against our community.

We now understand more or less how molecules can act as carriers of information and we have clarified some of the biochemical mech-anisms that enable us to read such information and react in conse-quence. We have become aware that smells are important aspects of our life, although not so essential as for other animals. Smells and tastes make life more pleasant—we can enjoy good food and relax in the resinous scents of the forest, or smell the fresh breeze at the seaside. We have also learned that smells can be used to manipulate the behaviour of animals, from insects to humans, in some cases with beneficial effects, but also with the potential of alarming misuse.

To return to the question of whether understanding the chemistry of olfaction has taken away that mysterious appeal from our olfactory experiences, as a scientist and in particular as a chemist, I feel even more attracted to this field now that I understand the molecular mechanism at the origin of our pleasures and emotions. This is particularly so when I discover that there is a common language widely spoken in nature at all levels of differentiation, from cross-talk between cells of an organism to the sweet words whispered by the boar to his love.

FURTHER READING

It is hoped that this walk across the world of smells, pheromones, and proteins may have piqued readers' interest and raised some further questions, prompting more curious readers to look deeper into some aspects, of which they had only a glimpse in this book. The scientific literature has become rich in information about the sense of the smell, but often papers published in specialized journals are addressed to readers with a specific background or at least with a solid scientific education. There follows a list of books ranging from chemistry and biochemistry to psychology, written in the style of 'popular science' accessible to anyone stimulated by scientific curiosity. Selected more technical reviews are also included for those who are familiar with molecules, genes, and proteins and who want to improve their scientific knowledge in the field of olfaction. In addition, the internet provides a rich resource where keywords such as smell, pheromone, olfaction, food flavour, or electronic nose, will bring up a large number of sites where smell is presented in a variety of contexts, from the smell of cities to olfactory events and exhibitions, often combining science and art, sometimes even crossing the blurred border between real science and imagination.

Pheromones

Tristram D. Wyatt. Pheromones and Animal Behaviour: Chemical Signals and Signatures. *2nd edn. Cambridge University Press, 2014.*

This is probably the best general book on pheromones. It takes a simple approach to the chemical language used by most animal species, but at the same time it is full of valuable scientific information. It is highly enjoyable while providing a mine of data for those working in the field.

Tristram D. Wyatt. 'The Search for Human Pheromones: The Lost Decades and the Necessity of Returning to First Principles'. Proc. R. Soc. B (2014), 282: 2994.

In this article, the author analyses the recent scientific literature on human pheromones. Navigating with a clear scientific mind through a foggy and muddy environment, he isolates the few experimental facts from myths, hypotheses, and imagination to conclude cautiously that, although not excluding the possibility that pheromonal communication might exist among humans, nevertheless the molecules mediating such behaviour still await discovery.

Bert Hölldobler and Edward O. Wilson. Journey to the Ants: A Story of Scientific Exploration. *Belknap Press, 1998.*

Bert Hölldobler and Edward O. Wilson. The Superorganism: The Beauty, Elegance, and Strangeness of Insect Societies. *W.W. Norton & Company, 2008.*

The authors of these two books (and several others, including a previous one which won them the Pulitzer Prize), have dedicated their lives to the study of ants, uncovering all the hidden and extraordinary aspects in the life of these small insects. Although not specifically focused on pheromones, these chemical messengers regulate most of behavioural aspects of insects, notably social species. The authors' prose, clear and captivating, make reading these books a pleasure, while learning a lot about science.

Perfumes and molecules

Perfumers were the first interest in the science of smell, and specifically in the olfactory properties of molecules. Chemistry and art are combined in the brains of these scientists, who pursue the dream of providing us with new olfactory emotions, both designing new molecules and mixing available fragrances as notes in a musical chord.

Charles S. Sell: Chemistry and the Sense of Smell. *Wiley, 2014.*

Charles Sell is a chemist working in industry and is interested in the correlations between molecular structure and odour. The book is focused on fragrances and is based on the long personal experience of the author in the field. Particularly interesting and informative are industrial aspects and applications of research in the chemistry of perfumes. Overall, the book provides a wealth of information on the chemistry of molecules that end up in our soaps, laundry, house cleaners and make our environment more agreable.

Charles S. Sell. 'On the unpredictability of odors'. Angewandte Chemie (2006), 45: 6254–61.

The thesis is that olfaction is so complex and multifaced that becomes almost impossible to design a new molecule with a desired smell.

Avery Gilbert. What the Nose Knows: The Science of Scent in Everyday Life. *Crown Publishing Group, 2015.*

A very well-written book, highly enjoyable and entertaining with a lot of anecdotes and curious facts. It is more focused on psychological and social aspects than on scientific information and is easily accessible to anyone.

Biochemistry, physiology, neurobiology

Gordon M. Shepherd. Neurogastronomy: How the Brain Creates Flavor and Why It Matters. *Columbia University Press, 2010.*

Gordon Shepherd is one of the pioneers in olfaction and before that, a leading neuroscientist. This book is focused on food flavour, as the title clearly indicates, and how our brain processes such important messengers of our everyday life.

Starting with the observation that the sense of smell is much more important in humans that previously suspected, Gordon Shepherd analyses how the smells of food are memorized as spatial patterns to construct olfactory images. Food preferences, craving, emotions, dieting, obesity, and even drug addiction are all considered in connection with the physiology of our sense of smell.

Anna Menini (ed.). The Neurobiology of Olfaction (Frontiers in Neuroscience). *CRC Press, 2009.*

This book presents various aspects of olfaction, both in vertebrates and in insects, ranging from peripheral events to brain activity produced by smells and neurogenesis of the olfactory system. It is a collection of studies written by leading scientists in each of the aspects covered.

Carla Mucignat-Caretta (ed.). Neurobiology of Chemical Communication (Frontiers in Neuroscience). *CRC Press, 2014.*

This book is more specifically focused on pheromonal communication, from chemistry to signal processing and behaviour, each chapter written by leading experts in the field of olfaction.

The following scientific publications are for readers with a good background in biochemistry, although who are not necessarily familiar with the science of smell.

L. Buck and R. Axel. 'A Novel Multigene Family May Encode Odorant Receptors: A Molecular Basis for Odor Recognition'. Cell (1991), 65: 175–87.

Nobel lectures of Linda Buck and Richard Axel, discoverer of olfactory receptors, are available at: http://www.nobelprize.org/nobel_prizes/medicine/laureates/2004/buck-lecture.html

http://www.nobelprize.org/nobel_prizes/medicine/laureates/2004/axel-lecture.html

P. J. Clyne, C. G. Warr, M. R. Freeman, D. Lessing, J. Kim, and J. R. Carlson. 'A Novel Family of Divergent Seven-Transmembrane Proteins: Candidate Odorant Receptors in Drosophila'. Neuron (1999), 22: 327–38.

S. Firestein. 'How the Olfactory System Makes Sense of Scents'. Nature (2001), 413: 211–18.

P. Mombaerts, F. Wang, C. Dulac, S. K. Chao, A. Nemes, M. Mendelsohn, J. Edmondson, and R. Axel. 'Visualizing an Olfactory Sensory Map'. Cell (1996), 87: 675–86.

P. A. Temussi. 'Sweet, Bitter and Umami Receptors: A Complex Relationship'. Trends Biochem. Sci. (2009), 34: 296–302.

Electronic noses

A.D. Wilson and M. Baietto. 'Applications and Advances in Electronic-Nose Technologies Sensors'. (2009), 9: 5099–148; doi:10.3390/s90705099.

This is a good review of the field of electronic noses—focusing on practical applications.

Krishna C. Persaud, Santiago Marco, and Agustin Gutierrez-Galvez (eds.). Neuromorphic Olfaction (Frontiers in Neuroengineering Series). CRC Press, Boca Raton, Florida, USA 2013.

This is an edited book with chapters written by experts in the field focusing on biomimetic aspects of artificial olfaction systems.

GLOSSARY OF MAIN CLASSES
OF CHEMICAL COMPOUNDS

Hydrocarbons—Contain only hydrogen and carbon in the molecule. Hydrocarbons can be saturated (the carbons are only connected to each other by single bonds), unsaturated (double or triple bonds can be present in the structure), or aromatic (particular arrangements of carbon atoms in a ring connected by bonds halfway between single and double). All aromatic hydrocarbons are planar and highly stable. A typical aromatic hydrocarbon is benzene. Such considerations on hydrocarbons apply to all other derivatives, where the reference hydrocarbon is considered as the skeleton of the molecules, to which groups and other atoms are attached.

Alcohols—Contain a hydroxyl group (–OH) in the molecule. This group can interact with water and confers water solubility to alcohols, highest in the smallest members of the family, such as methanol and ethanol. Lower alcohols are produced during the fermentation of foods, others are components of the scent of grass, flowers, and mushrooms.

Aldehydes and ketones—Are characterized by a carbon-oxygen double bond, present at the end of the hydrocarbon chain or in the middle, respectively. Aldehydes and ketones are less soluble in water and more volatile than their corresponding alcohols. They include a large number of compounds with pleasant smells, occurring both in flowers and in foods.

Carboxylic acids—The functional group of organic acid is –COOH, a carbon linked to an oxygen with a double bond and to an –OH group with a single bond. They present strong acid reaction and are found in a variety of foods, such as acetic acid in vinegar, citric in lemon, malic in apples, and tartaric in grapes and wines.

Esters—Can be regarded as (and can be prepared by) the combination of a carboxylic acid and an alcohol with a loss of a water molecule in the process. Can also be ideally derived from a carboxylic acid by replacing the oxygen-hydrogen bond with an oxygen-carbon bond. Common examples of esters are both animal fats and vegetable oils, products of condensation between a molecule of glycerol and three molecules of caboxylic acids. Esters are endowed with fruity smells and occur widely in all varieties of fruits. Several insect peromones are esters.

Lactones—These molecules are esters where the original alcoholic and carboxylic groups belonged to the same molecules. Therefore, their reaction produces cyclic structures. Lactones are found among food aroma components, fruits, and insect pheromones.

Amines—Are characterized by a nitrogen atom attached to one, two, or three carbon chains. They are the product of degradation of proteins and act with their repulsive odour as warning signals for decomposing food.

Amides—They can be ideally built by replacing the hydroxyl group of carboxylic acids with an amino group. Proteins are polyamides, resulting from the condensation of amino acids.

Amino acid—Contain both an amino group and a carboxylic group and constitute the building blocks of proteins.

NOTES AND REFERENCES

PREFACE

1. John E Amoore: *Molecular Basis of Odor*. (American lecture series, publication no. 773. A monograph in the Bannerstone division of American lectures in living chemistry). Thomas, 1970; J. E. Amoore, J. W. Johnston Jr., and M. Rubin: 'The Sterochemical Theory of Odor'. *Scientific American* (1964), 210: 42–9; J. E. Amoore: 'Specific Anosmia: A Clue to the Olfactory Code'. *Nature* (1967), 214: 1095–8.
2. P. P. Graziadei and G. A. Monti Graziadei: 'Neurogenesis and Plasticity of the Olfactory Sensory Neurons'. *Ann NY Acad. Sci.* (1985), 457:127–42.
3. Konrad Lorenz: *King Solomon's Ring*, 2nd edn. Routledge Classics (Taylor & Francis Group), 2002.

CHAPTERS

1. Patrick Suskind: *Perfume: The Story of a Murderer*. Vintage, 2001.
2. Amoore: *Molecular Basis of Odor*; Moore, Johnston, and Rubin: 'The Sterochemical Theory of Odor'; Amoore: 'Specific Anosmia: A Clue to the Olfactory Code'.
3. J. E. Amoore: 'Specific Anosmia and the Concept of Primary Odors'. *Chem Senses* (1977), 2: 267–81.
4. To be precise, there are two isomers of androstenol, called α and β. One has an odour similar to that of androstenone, the other is mainly musk with a faint urinous note.
5. C. Sell: 'On the Unpredictability of Odors'. *Angewandte Chemie* (2006), 45: 6254–61.
6. See Gunther Ohloff, Wilhelm Pickenhagen, and Philip Kraft: *Scent and Chemistry*. Wiley, 2011; D. H. Pybus and C. S. Sell: *The Chemistry of Fragrances*. Royal Society of Chemistry, London, 2004; Charles Sell: *Understanding Fragrance Chemistry*. Allured Publishing Corporation, 2008.
7. Steffan Arctander: *Perfume and Flavor Chemicals (Aroma Chemicals)* Two-Book Set. Allured Publishing Corporation, 2000.
8. David H. Hubel: *Eye, Brain, and Vision*. Scientific American Library, 1988; Jeremy Nathans, Darcy Thomas, and David S. Hogness: 'Molecular Genetics

of Human Color Vision: The Genes Encoding Blue, Green, and Red Pigments'. *Science* (1986), 232: 193–202.

9. Amoore, Johnston, and Rubin: 'The Sterochemical Theory of Odor'; Amoore: 'Specific Anosmia: A Clue to the Olfactory Code'.

10. K.-E. Kaissling: 'Chemo-Electrical Transduction in Insect Olfactory Receptors'. *Annual Review of Neurosciences* (1986), 9: 21–45; K.-E. Kaissling: *R. H. Wright Lectures on Insect Olfaction*, Ed. K. Colbow. Simon Fraser University, Burnaby, B.C., Canada (1987).

11. A. Butenandt, R. Beckamnn, and E. Hecker: 'Über den Sexuallockstoff des Seidenspinners. 1. Der biologische Test und die Isolierung des reinen Sexuallockstoffes Bombykol'. *Hoppe-Seylers Zeitschrift für Physiologische Chemie* (1961), 324: 71.

12. Tristram D. Wyatt: *Pheromones and Animal Behavior: Chemical Signals and Signatures*. 2nd edn. Cambridge University Press, 2014.

13. The notation using L and D is the traditional one and still used sometimes for sugars and amino acids. More correctly, and less confusing, the modern notation uses the symbols (R) and (S), to label the chirality of single asymmetric centres (most often a carbon linked to four different groups) according to a standard nomenclature. Finally, the symbols (+) and (-) refer to the experimentally measured rotation of the plane of polarized light (right and left, respectively).

14. The Pherobase: Database of Insect Pheromones and Semiochemicals. Available at http//www.pherobase.net/.

15. John Brand Free: *Pheromones of Social Bees*. Cornell University Press, 1987.

16. Bert Hölldobler and Edward O. Wilson: *The Superorganism: The Beauty, Elegance, and Strangeness of Insect Societies*. W. W. Norton & Company, 2008; Bert Hölldobler and Edward O. Wilson: *Journey to the Ants: A Story of Scientific Exploration*. Belknap Press, 1998.

17. Keith N. Slessor, Lori-Ann Kaminski, G. G. S. King, John H. Borden, and Mark L. Winston: 'Semiochemical Basis of the Retinue Response to Queen Honey Bees'. *Nature* (1988), 332: 354–6.

18. Karl von Frisch: *The Dance Language and Orientation of Bees*. The Belknap Press of Harvard University Press, 1967.

19. W. M. Farina, C. Grüter, and P. C. Díaz: 'Social Learning of Floral Odours Inside the Honeybee Hive'. *Proc Biol Sci.* (2005) 272: 1923–8.

20. More information on mosquitoes can be found on the American Mosquito Control Association's website at http://www.mosquito.org.

21. A. M. Pohlit, N. P. Lopes, R. A. Gama, W. P. Tadei, and V. F. Neto: 'Patent Literature on Mosquito Repellent Inventions which Contain Plant Essential Oils—A Review'. *Planta Med.* (2011) 77: 598–617. doi: 10.1055/s-0030-1270723.

22. A very recent account on pheromones both in vertebrates and in insects can be found in C. Mucignat-Caretta (ed.): *Neurobiology of Chemical Communication*. CRC Press, 2014.

23. Richard L. Doty: *The Great Pheromone Myth*. Johns Hopkins University Press, 2010.

24. J. L. Hurst and R. J. Beynon: 'Scent Wars: The Chemobiology of Competitive Signalling in Mice. *Bioessays* (2004), 26: 1288–98.

25. B. Schaal, G. Coureaud, D. Langlois, C. Giniès, E. Sémon, and G. Perrier: Chemical and Behavioural Characterization of the Rabbit Mammary Pheromone. *Nature* (2003), 424: 68–72.

26. B. Schaal: 'Pheromones for Newborns'. In C. Mucignat-Caretta (ed.) *Neurobiology of Chemical Communication*. CRC Press, 2014, Chapter 17.

27. L. Buck and R. Axel: 'A Novel Multigene Family May Encode Odorant Receptors: A Molecular Basis for Odor Recognition'. *Cell* (1991), 65: 175–87.

28. For those interested in receptors and neurotransmitters, a large choice of books is available from popular science to more technical guides. See, among others, Solomon H. Snyder: *Drugs and the Brain* (A Scientific American Library paperback) W H Freeman & Co, 1996; Solomon H. Snyder: *Science and Psychiatry: Ground-breaking Discoveries in Molecular Neuroscience*. American Psychiatric Publishing, 2008; E. C. Hulme: *Receptor Biochemistry: A Practical Approach* (Practical Approach Series) IRL Press, 1990.

29. A useful repertoire of olfactory thresholds is: Zeist L. J. van Gemert: *Flavour Thresholds. Compilations of Flavour Threshold Values in Water and Other Media*. 2nd enlarged and revised edn. Oliemans Punter and Partners BV, 2001.

30. Amoore, Pelosi, and Forrester: 'Specific Anosmia to 5a-androst-16-en-3-one and ω-pentadecalactone: The Urinous and Musky Primary Odors'. *Chem. Sens.* (1977), 2, 401.

31. A. Keller, H. Zhuang, Q. Chi, L. B. Vosshall, and H. Matsunami: 'Genetic Variation in a Human Odorant Receptor Alters Odour Perception'. *Nature* (2007), 449, 468–73.

32. P. Pelosi, N. E. Baldaccini, A. M. Pisanelli: 'Identification of a Specific Olfactory Receptor for 2-isobutyl-3-methoxypyrazine'. *Biochem. J.* (1982), 201, 245–8.

33. Thomas Welton: 'Room-Temperature Ionic Liquids'. *Chem. Rev.* (1999), 99: 2071–84. doi:10.1021/cr980032t; Michael Freemantle, Tom Welton, and Robin D Rogers: *An Introduction to Ionic Liquids*. Royal Society of Chemistry, 2009.

34. E. Bignetti, A. Cavaggioni, P. Pelosi, K. C. Persaud, R. T. Sorbi, and R. Tirindelli: 'Purification and Characterization of an Odorant Binding Protein from Cow Nasal Tissue'. *Eur. J. Biochem.* (1985), 149: 227–31.

35. J. Pevsner, R. R. Trifiletti, S. M. Strittmatter, and S. H. Snyder: 'Isolation and Characterization of an Olfactory Receptor Protein for Odorant Pyrazines'. *Proc. Natl. Acad. Sci. USA* (1985), 82: 3050–4.

36. R. G. Vogt and L. M. Riddiford: 'Pheromone Binding and Inactivation by Moth Antennae'. *Nature* (1981), 293:161–3.

37. M. A. Bianchet, G. Bains, P. Pelosi, J. Pevsner, S. H. Snyder, et al.: 'The Three Dimensional Structure of Bovine Odorant-Binding Protein and Its

Mechanism of Odor Recognition'. *Nat. Struct. Biol.* (1996), 3: 934–9; M. Tegoni, R. Ramoni, E. Bignetti, S. Spinelli, and C. Cambillau: 'Domain Swapping Creates a Third Putative Combining Site in Bovine Odorant Binding Protein Dimer'. *Nat. Struct. Biol.* (1996), 3: 863–7.

38. B. H. Sandler, L. Nikonova, W. S. Leal, and J. Clardy: 'Sexual Attraction in the Silkworm Moth: Structure of the Pheromone-Binding-Protein-Bombykol Complex'. *Chem. Biol* (2000), 7:143–51.

39. C. Mucignat-Caretta, A. Caretta, and A. Cavaggioni: 'Acceleration of Puberty Onset in Female Mice by Male Urinary Proteins'. *J. Physiol.* (1995), 486: 517–22.

40. Mucignat-Caretta: *Neurobiology of Chemical Communication*; P. Pelosi: 'Odorant-Binding Proteins'. *Crit. Rev. Biochem. Mol. Biol.* (1994), 29: 199–228; Gary J. Blomquist and Richard G. Vogt (eds.): *Insect Pheromone Biochemistry and Molecular Biology: The Biosynthesis and Detection of Pheromones and Plant Volatiles.* Elsevier Academic Press, 2003; P. Pelosi, J. J. Zhou, L. P. Ban, and M. Calvello: 'Soluble Proteins in Insect Chemical Communication'. *Cell Mol. Life Sci.* (2006), 63: 1658–1676; W. S. Leal: 'Odorant Reception in Insects: Roles of Receptors, Binding Proteins, and Degrading Enzymes'. *Ann. Rev. Entomol.* (2013), 58: 373–91; P. Pelosi, I. Iovinella, A. Felicioli, and F. R. Dani: 'Soluble Proteins of Chemical Communication: An Overview Across Arthropods'. *Front Physiol.* (2014), 5:320. doi: 10.3389/fphys.2014.00320. eCollection 2014.

41. Buck and Axel: 'A Novel Multigene Family May Encode Odorant Receptors', Cell. 65: 175–87.

42. You can read the Nobel lectures of Linda Buck and Richard Axel on the website: http://www.nobelprize.org/nobel_prizes/medicine/laureates/2004/buck-lecture.html and http://www.nobelprize.org/nobel_prizes/medicine/laureates/2004/axel-lecture.html.

43. Piali Sengupta: 'Generation and Modulation of Chemosensory Behaviors in *C. elegans*' *Pflugers Arch—Eur. J. Physiol.* (2007), 454: 721–34.

44. Excellent studies on olfactory receptors and the transduction of odours have been published in the last few years, most of which are easily accessible by the non-specialist. See P. Mombaerts: 'Molecular Biology of Odorant Receptors in Vertebrates'. *Ann. Rev. Neurosci.* (1999), 22: 487–509; Anna Menini (ed.): *The Neurobiology of Olfaction (Frontiers in Neuroscience)* CRC Press, 2009; C. I. Bargmann: 'Comparative Chemosensation from Receptors to Ecology'. *Nature* (2006), 444(7117): 295–301; S. Firestein: 'How the Olfactory System Makes Sense of Scents'. *Nature* (2001), 413: 211–18; J. G. Hildebrand and G. M. Shepherd: 'Mechanisms of Olfactory Discrimination: Converging Evidence for Common Principles Across Phyla'. *Ann. Rev. Neurosci.* (1997), 20: 595–631.

45. For more detailed information on the vomeronasal system and its receptors, see R. Tirindelli, C. Mucignat-Caretta, and N. J. Ryba: 'Molecular

Aspects of Pheromonal Communication Via the Vomeronasal Organ of Mammals'. *Trends Neurosci.* (1998), 11: 482–6; D. Trotier: 'Vomeronasal Organ and Human Pheromones'. *European Annals of Otorhinolaryngology, Head and Neck Diseases* (2011), 128: 184–190; X. Ximena Ibarra-Soria, M. O. Levitin, and D. W. Logan: 'The Genomic Basis of Vomeronasal-mediated Behaviour'. *Mamm. Genome* (2014), 25: 75–86.

46. S. A. Roberts, D. M. Simpson, S. D. Armstrong, A. J. Davidson, D. H. Robertson, L. McLean, R. J. Beynon, and J. L. Hurst: 'Darcin: A Male Pheromone that Stimulates Female Memory and Sexual Attraction to an Individual Male's Odour'. *BMC Biology* (2010), 8: 75.

47. S. C. Kinnamon: 'Neurosensory Transmission Without a Synapse: New Perspectives on Taste Signaling'. *BMC Biol.* (2013), 11: 42. doi: 10.1186/1741-7007-11-42; P. A. Temussi: 'Sweet, Bitter and Umami Receptors: A Complex Relationship'. *Trends Biochem. Sci.* (2009), 34: 296–302; A. A. Bachmanov and G. K. Beauchamp: 'Taste Receptor Genes'. *Ann. Rev. Nutr.* (2007), 27: 389–414.

48. M. Parmentier, F. Libert, S. Schurmans, S. Schiffmann, A. Lefort, D. Eggerickx, C. Ledent, C. Mollereau, C. Gérard, J. Perret, A. Grootegoed, and G. Vassart: 'Expression of Members of the Putative Olfactory Receptor Gene Family in Mammalian Germ Cells'. *Nature* (1992), 355: 453–5.

49. M. Spehr, G. Gisselmann, A. Poplawski, J. A. Riffell, C. H. Wetzel, R. K. Zimmer, and H. Hatt: 'Identification of a Testicular Odorant Receptor Mediating Human Sperm Chemotaxis'. *Science* (2003), 299: 2054–8.

50. These studies summarize much of the data accumulated in recent years on the links between olfactory receptors and cancer: NaNa Kang and JaeHyung Koo: 'Olfactory Receptors in Non-chemosensory Tissues'. *BMB Reports* (2012), 45: 612–22; Simon R. Foster, Eugeni Roura, and Walter G. Thomas: 'Extrasensory Perception: Odorant and Taste Receptors Beyond the Nose and Mouth'. *Pharmacology & Therapeutics* (2014), 142: 41–61.

51. D. Busse, P. Kudella, N.-M. Gruning, G. Gisselmann, S. Stander, T. Luger, F. Jacobsen, L. Steinstraer, R. Paus, P. Gkogkolou, M. Bohm, H. Hatt, and H. Benecke: 'A Synthetic Sandalwood Odorant Induces Wound Healing Processes in Human Keratinocytes via the Olfactory Receptor OR2AT4'. *Journal of Investigative Dermatology* (2014), 134: 2823–32.

52. P. J. Clyne, C. G. Warr, M. R. Freeman, D. Lessing, J. Kim, and J. R. Carlson: 'A Novel Family of Divergent Seven-Transmembrane Proteins: Candidate Odorant Receptors in *Drosophila*'. *Neuron* (1999), 22: 327–338; L. B. Vosshall, H. Amrein, P. S. Morozov, A. Rzhetsky, and R. Axel: 'A Spatial Map of Olfactory Receptor Expression in the *Drosophila* Antenna'. *Cell* (1999), 96: 725–736.

53. P. Mombaerts, F. Wang, C. Dulac, S. K. Chao, A. Nemes, M. Mendelsohn, J. Edmondson, and R. Axel: 'Visualizing an Olfactory Sensory Map'. *Cell* (1996), 87: 675–86.

54. C. G. Galizia, S. Sachse, A. Rappert, and R. Menzel: 'The Glomerular Code for Odor Representation is Species Specific in the Honeybee *Apis mellifera*'. *Nat. Neurosci.* (1999), 2: 473–8.

55. More detailed information on facts and ideas about human pheromones can be found in Tristram Wyatt: TEDx Talk: Smelly Mystery of Human Pheromones (available at: http://www.ted.com/talks/tristram_wyatt_the_smelly_mystery_of_the_human_pheromone) and in Doty: *The Great Pheromone Myth*.

56. Schaal, 'Pheromones for Newborns'. In Mucignat-Caretta C., (ed.): *Neurobiology of Chemical Communication*. Boca Raton (FL): CRC Press; 2014. Chapter 17.

57. K. Persaud and G. Dodd: 'Analysis of Discrimination Mechanisms in the Mammalian Olfactory System Using a Model Nose'. *Nature* (1982): 299: 352–5.

58. K. C. Persaud and P. Pelosi: 'An Approach to an Artificial Nose'. *Trans. Amer. Soc. Artif. Int. Organs* (1985), 31(1): 297–300; Krishna C. Persaud: 'Biomimetic Olfactory Sensors'. *IEEE Sensors Journal* (2012), 12: 3108–12; M. Bernabei, K. C. Persaud, S. Pantalei, E. Zampetti, and R. Beccherelli: 'Large-Scale Chemical Sensor Array Testing Biological Olfaction Concepts'. *IEEE Sensors Journal* (2012), 12: 3174–83; P. Pelosi, R. Mastrogiacomo, I. Iovinella, E. Tuccori, and K. C. Persaud: 'Structure and Biotechnological Applications of Odorant-binding Proteins'. *Appl. Microbiol. Biotechnol.* (2013), 98: 61–70; K. C. Persaud, A. M. Pisanelli, S. Szyszko, M. Reichl, G. Horner, W. Rakow, H. J. Keding, and H. Wessels: 'A Smart Gas Sensor for Monitoring Environmental Changes in Closed Systems: Results from the MIR Space Station. Sensors and Actuators B'. *Chemical* (1999), 55: 118–26.

59. David Dosa: *Making Rounds with Oscar: The Extraordinary Gift of an Ordinary Cat.* Hyperion, 2011.

INDEX

BAD MOVES

How Decision Making Goes Wrong, and the Ethics of Smart Drugs

Barbara Sahakian and Jamie Nicole LaBuzetta

BARBARA J. SAHAKIAN &
JAMIE NICOLE LABUZETTA

978-0-19-966848-9 | Paperback | £8.99

"With this accessible primer, full of medical anecdotes and clear explanations, Sahakian and Labuzetta prepare the public for an informed discussion about the role of drugs in our society."   **Nature**

The realization that smart drugs can improve cognitive abilities in healthy people has led to growing general use, with drugs easily available via the Internet. Sahakian and Labuzetta raise ethical questions about the availability of these drugs for cognitive enhancement, in the hope of informing public debate about an increasingly important issue.

CURIOUS TALES FROM CHEMISTRY

The Last Alchemist in Paris and Other Episodes

Lars Öhrström

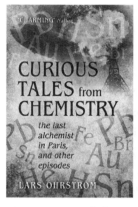

978-0-19-874392-7 | Paperback | £11.99

"Chemistry is a messy business. Öhrström...
writes very well indeed, to remind us that
physical chemistry is everywhere and can
explain almost every material thing."

The Guardian

"[A] charming mish mash of a primer."

Nature

Lars Öhrström introduces us to a variety of
elements from S to Pb through tales of ordinary
and extraordinary people from around the globe.
We meet African dictators controlling vital supplies
of uranium, and eighteenth-century explorers
searching out sources of precious metals. We find
out why the Hindenburg airship was tragically filled
with hydrogen, not helium; and why nail-varnish
remover played a key part in World War I. In each
chapter, we find out about the distinctive properties
of each element and the concepts and principles that
have enabled scientists to put it to practical use.
These are the fascinating (and sometimes terrifying)
stories of chemistry in action.

REACTIONS

The Private Life of Atoms

Peter Atkins

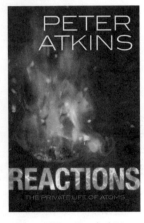

978-0-19-966880-9 | Paperback | £12.99

"The perfect antidote to science phobia."

Booklist

Peter Atkins captures the heart of chemistry in this book, through an innovative, closely integrated design of images and text, and his characteristically clear, precise, and economical exposition. Explaining the processes involved in chemical reactions, he begins by introducing a 'tool kit' of basic reactions, such as precipitation, corrosion, and catalysis, and concludes by showing how these building blocks are brought together in more complex processes such as photosynthesis, to provide a concise and intellectually rewarding introduction to the private life of atoms.

TESTOSTERONE

Sex, Power, and the Will to Win

Joe Herbert

978-0-19-872497-1 | Hardback | £16.99

"It is the best of hormones; it is the worst of hormones. Joe Herbert leads a guided tour through human evolution using the multifaceted hormone as his lens and vehicle."

New Scientist

Testosterone underlies the activation of masculinity: it changes the body and brain to make a male. It is involved not only in sexuality but in driving aggression, competitiveness, risk-taking—all elements that were needed for successful survival and reproduction in the past. In *Testosterone*, Joe Herbert explains the nature of this potent hormone, how it operates, what we know about its role in influencing various aspects of behaviour in men, and what we are beginning to understand of its role in women. From rape to gang warfare among youths, understanding the workings of testosterone is critical to enable us to manage its continuing powerful effects in modern society.

Sign up to our quarterly e-newsletter **http://academic-preferences.oup.com/**

WHAT IS CHEMISTRY?

Peter Atkins

978-0-19-968398-7 | Hardback | £11.99

"Atkins wins his readers' attention simply through an elegant and lucid description of the subject he loves." **Nature**

In *What is Chemistry?* Peter Atkins encourages us to look at chemistry anew, through a chemist's eyes, to understand its central concepts and to see how it contributes not only towards our material comfort, but also to human culture. He shows how chemistry provides the infrastructure of our world, through the chemical industry, the fuels of heating, power generation, and transport, as well as the fabrics of our clothing and furnishings. By considering the remarkable achievements that chemistry has made, and examining its place between both physics and biology, Atkins presents a fascinating, clear, and rigorous exploration of the world of chemistry.

WHAT IS LIFE?

How Chemistry Becomes Biology

Addy Pross

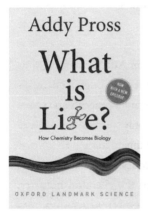

978-0-19-878479-1 | Paperback | £9.99

"Pross does an excellent job of succinctly conveying the difficulty in crafting an unambiguous general definition of life and provides a road map to much of the work on the origin of life done by chemists in the past 50 years. The book is worth the read for these discussions alone." *Chemical Heritage*

Living things are hugely complex and have unique properties, such as self-maintenance and apparently purposeful behaviour which we do not see in inert matter. So how does chemistry give rise to biology? What could have led the first replicating molecules up such a path? Now, developments in the emerging field of 'systems chemistry' are unlocking the problem. The gulf between biology and the physical sciences is finally becoming bridged.

Sign up to our quarterly e-newsletter http://academic-preferences.oup.com/